Social Semantic Web Mining

Synthesis Lectures on the Semantic Web: Theory and Technology

Editors
Ying Ding, *Indiana University*
Paul Groth, *VU University Amsterdam*

Founding Editor Emeritus
James Hendler, *Rensselaer Polytechnic Institute*

Synthesis Lectures on the Semantic Web: Theory and Application is edited by Ying Ding of Indiana University and Paul Groth of VU University Amsterdam. Whether you call it the Semantic Web, Linked Data, or Web 3.0, a new generation of Web technologies is offering major advances in the evolution of the World Wide Web. As the first generation of this technology transitions out of the laboratory, new research is exploring how the growing Web of Data will change our world. While topics such as ontology-building and logics remain vital, new areas such as the use of semantics in Web search, the linking and use of open data on the Web, and future applications that will be supported by these technologies are becoming important research areas in their own right. Whether they be scientists, engineers or practitioners, Web users increasingly need to understand not just the new technologies of the Semantic Web, but to understand the principles by which those technologies work, and the best practices for assembling systems that integrate the different languages, resources, and functionalities that will be important in keeping the Web the rapidly expanding, and constantly changing, information space that has changed our lives.
Topics to be included:

- Semantic Web Principles from linked-data to ontology design

- Key Semantic Web technologies and algorithms

- Semantic Search and language technologies

- The Emerging "Web of Data" and its use in industry, government and university applications

- Trust, Social networking and collaboration technologies for the Semantic Web

- The economics of Semantic Web application adoption and use

- Publishing and Science on the Semantic Web

- Semantic Web in health care and life sciences

Social Semantic Web Mining

Tope Omitola, Sebastián A. Ríos, and John G. Breslin

ISBN: 978-3-031-79458-2 paperback
ISBN: 978-3-031-79459-9 ebook

DOI 10.1007/978-3-031-79459-9

A Publication in the Springer series
SYNTHESIS LECTURES ON THE SEMANTIC WEB: THEORY AND TECHNOLOGY

Lecture #10
Series Editors: Ying Ding, *Indiana University*
      Paul Groth, *VU University Amsterdam*
Founding Editor Emeritus: James Hendler, *Rensselaer Polytechnic Institute*
Series ISSN
Print 2160-4711 Electronic 2160-472X

Social Semantic Web Mining

Tope Omitola
University of Southampton

Sebastián A. Ríos
University of Chile

John G. Breslin
National University of Ireland Galway

SYNTHESIS LECTURES ON THE SEMANTIC WEB: THEORY AND TECHNOLOGY #10

ABSTRACT

The past ten years have seen a rapid growth in the numbers of people signing up to use Web-based social networks (hundreds of millions of new members are now joining the main services each year) with a large amount of content being shared on these networks (tens of billions of content items are shared each month). With this growth in usage and data being generated, there are many opportunities to discover the knowledge that is often inherent but somewhat hidden in these networks. Web mining techniques are being used to derive this hidden knowledge. In addition, the Semantic Web, including the Linked Data initiative to connect previously disconnected datasets, is making it possible to connect data from across various social spaces through common representations and agreed upon terms for people, content items, etc.

In this book, we detail some current research being carried out to semantically represent the implicit and explicit structures on the Social Web, along with the techniques being used to elicit relevant knowledge from these structures, and we present the mechanisms that can be used to intelligently mesh these semantic representations with intelligent knowledge discovery processes. We begin this book with an overview of the origins of the Web, and then show how web intelligence can be derived from a combination of web and Social Web mining. We give an overview of the Social and Semantic Webs, followed by a description of the combined Social Semantic Web (along with some of the possibilities it affords), and the various semantic representation formats for the data created in social networks and on social media sites.

Provenance and provenance mining is an important aspect here, especially when data is combined from multiple services. We will expand on the subject of provenance and especially its importance in relation to social data. We will describe extensions to social semantic vocabularies specifically designed for community mining purposes (SIOCM). In the last three chapters, we describe how the combination of web intelligence and social semantic data can be used to derive knowledge from the Social Web, starting at the community level (macro), and then moving through group mining (meso) to user profile mining (micro).

KEYWORDS

World Wide Web, Web Mining, Web Intelligence, Semantic Web, Social Web, Social Semantic Web, Provenance, Knowledge Discovery, Knowledge Management

We would like to dedicate this book to our families,
who have supported us and been patient with us throughout
the writing of this book.

Contents

Acknowledgments

Tope Omitola would like to express gratitude for the managerial support of Professor Sir Nigel Shadbolt of the Web and Internet Science Group, University of Southampton and of Dr. John Davies of British Telecommunications during the writing of this book. He would also like to acknowledge the support of EPSRC (Grant Number EP/K503770/1) during this time.

Sebastián Ríos would like to thank the continuous support from the Complex Engineering Systems Institute (ICM: P-05-004-F, CONICYT: FBO16) and the research grant for Initiation into Research (FONDECYT), project code 11090188, entitled "Semantic Web Mining Techniques to Study Enhancements of Virtual Communities."

John Breslin would like to acknowledge input from his collaborators Dr. Fabrizio Orlandi and Dr. Alexandre Passant. He would also like to acknowledge that his work on this publication has emanated from research supported in part by research grants from Science Foundation Ireland (SFI) under Grant Number SFI/08/CE/I1380 (DERI Líon 2) and also under Grant Number SFI/12/RC/2289 (Insight).

Tope Omitola, Sebastián A. Ríos, and John G. Breslin
January 2015

Grant Aid

This publication was grant-aided by the Publications Fund of
National University of Ireland Galway

Rinneadh maoiniú ar an bhfoilseacháin seo trí Chiste Foilseacháin
Ollscoil na hÉireann Gaillimh

CHAPTER 1

Introduction and the Web

1.1 INTRODUCTION

The past ten years have seen a rapid growth in the numbers of people signing up to use Web-based social networks (hundreds of millions of new members are now joining the main services each year) with a large amount of content being shared on these networks (tens of billions of content items are shared each month).

With this growth in usage and data being generated, there are many opportunities to discover the knowledge that is often inherent but somewhat hidden in these networks. Web mining techniques are being used to derive this hidden knowledge.

On the other hand, the Semantic Web, including the Linked Data initiative to connect previously disconnected datasets, is making it possible to connect data from across various social spaces through common representations and agreed-upon terms for people, content items, etc.

In this book, we detail some current research being carried out to semantically represent the implicit and explicit structures on the Social Web, along with the techniques being used to elicit relevant knowledge from these structures, and we present the mechanisms that can be used to intelligently mesh these semantic representations with intelligent knowledge discovery processes.

We will begin this book with an overview of the origins of the Web (Chapter 1), and then show how web intelligence can be derived from a combination of web and Social Web mining (Chapter 2). We give an overview of the Social and Semantic Webs (Chapters 3 and 4), followed by a description of the combined Social Semantic Web (along with some of the possibilities it affords) (Chapter 5), and the various semantic representation formats for the data created in social networks and on social media sites.

Provenance and provenance mining is an important aspect here, especially when data is combined from multiple services. We will expand on the subject of provenance and especially its importance in relation to social data (Chapter 6).

We will describe extensions to social semantic vocabularies specifically designed for community mining purposes (SIOCM).

In the last three chapters, we describe how the combination of web intelligence and social semantic data can be used to derive knowledge from the Social Web, starting at the community level (macro), and then moving through group mining (meso) to user profile mining (micro).

1.2   THE WORLD WIDE WEB

"The World Wide Web ("WWW" or simply the "Web") is a global, read-write information space. Text documents, images, multimedia and many other items of information, referred to as resources, are identified by short, unique, global identifiers called Uniform Resource Identifiers (URIs) so that each can be found, accessed and cross-referenced in the simplest possible way."[1]

As the reader will note, the Web is not a synonym for the Internet. The Internet refers to the physical network and protocols, while the Web refers to a framework running on top of it. The public debut of the Web was on August 6, 1991, when Tim Berners-Lee (one of its creators) posted a short summary of the World Wide Web project on the alt.hypertext newsgroup.[2] Berners-Lee's breakthrough was to combine hypertext with the Internet in an easy-to-deploy framework, that went well beyond previous applications like Gopher.

1.2.1   HISTORY AND EVOLUTION OF THE WEB

In 1980, Tim Berners-Lee wrote some software (ENQUIRE) which allowed one to browse through "cards" about resources at CERN (the European Organization for Nuclear Research) via bidirectional hyperlinks. By 1990, Berners-Lee was working on a GUI (graphical user interface) to perform browsing of hypertext, naming it the "World Wide Web" [Chakrabarti, 2002], and the Web's operation began in the following year, as detailed above.

In 1992, other programs were developed to browse hypertext on the Web such as Viola, Erwise, Midas, and Cello. By early 1993, the first version of Mosaic was completed by Mark Andreessen at the NCSA (National Center for Supercomputer Applications). Simultaneously, CERN developed a new improved HTML protocol (HTTP) to send HTML (Hypertext Markup Language) documents and other data over the growing Internet. To do so, they developed a server called CERN httpd (Hyper Text Transfer Protocol Daemon).

When it was announced that Gopher was no longer free to use, CERN announced that the World Wide Web would be free to anyone, with no fees, this producing a rapid shift away from Gopher and toward the Web. This resulted in an exponential growth in traffic on the Web and also in website development (detailed information can be found in Chakrabarti [2002]).

By October 1994, *"Sir Tim Berners-Lee, inventor of the World Wide Web, left the European Organization for Nuclear Research (CERN) and founded the World Wide Web Consortium (W3C) at the Massachusetts Institute of Technology Laboratory for Computer Science (MIT/LCS) with support from the Defense Advanced Research Projects Agency (DARPA), which had pioneered the Internet, and the European Commission."*[3] In the same year, the Mosaic Communications Corporation was also created. Later, Mosaic changed its name to the more well known Netscape Communications Corporation.

---

[1] http://en.wikipedia.org/wiki/World_Wide_Web.
[2] https://groups.google.com/forum/#!msg/alt.hypertext/eCTkkOoWTAY/bJGhZyooXzkJ.
[3] http://en.wikipedia.org/wiki/W3C.

To summarize, from a technical point of view the three key features that allow the existence of the Web are:

- Uniform Resource Identifier (URI), which is a universal system for referencing resources on the Web, such as webpages, images, videos, etc.

- Hypertext Transfer Protocol (HTTP), which specifies how the client (browser) and server communicate with each other.

- Hypertext Markup Language (HTML), used to define the structure and content of hypertext documents (webpages).

A term often conflated with URI is URL (Uniform Resource Locator). A URL is a type of URI that not only identifies a resource but allows one to locate that resource at the address that the URI points to.

1.2.2 A SHORT EXPLANATION OF HOW THE WEB WORKS

The Web is based on the simple but powerful idea of browsing a hyperlinked collections of hypertext documents [Morisseau-Leroy et al., 2001]. Usually, a blade/rack/tower server running some server software for the Web (Apache, Caucho Resin, PWS, etc.) is called the "web server." However, the computer can be any machine, even a desktop computer or a low-cost Raspberry Pi microcomputer.

It is important to differentiate the hardware server from the server software. The same holds for the client computer of the person who is trying to browse the Web (see Fig. 1.1). The software for browsing hypertext documents on the Web (Chrome, Firefox, Safari, Opera, Internet Explorer, etc.) is called the client software, whereas the actual computer is the hardware client. This is why in Fig. 1.1, we show client and server software which is running on different computers.

Let us imagine a person who is at home and is trying to access the website for the University of Tokyo. To do so, they will have some browser software installed on their computer, for example, Firefox. To begin with, they could type in the URL (if known) for the University of Tokyo's main page in English[4] and then press "enter." Alternatively they could use a search engine to find and click on the web address. At this point, the browser generates an "HTTP request" (back arrow on Fig. 1.1). This request is then routed over many different networks (composed of routers, switches, firewalls, satellite connections, etc.) until it reaches the server whose address corresponds to the desired one.[5] Once that happens, the server (software), for example, Apache, resolves the request and sends the requested page back to the client. The webpage (in this case "index_e.html") is divided into small pieces of data called datagrams (or data packages) and these are sent back to the client over the Internet. At the client, these packages are reassembled to get the original data.

---

[4]http://www.u-tokyo.ac.jp/index_e.html.
[5]To reduce complexity and provide the reader with just an overview of how the Web works, we have avoided an explanation of datagram routing (IP tables, IP numbers, DNS, Internet registries, etc.).

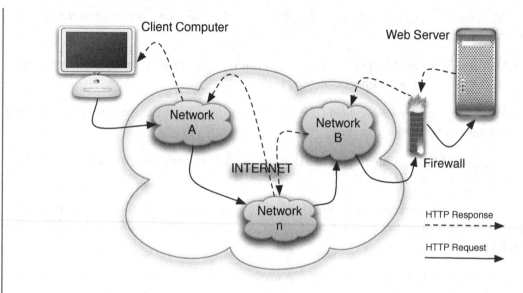

Figure 1.1: World Wide Web requests over the Internet.

In this way, the person who is using the client software is able to read a copy of the webpage "index_e.html" which was originally stored on the web server "www.u-tokyo.ac.jp."

1.3   GLOBAL IMPACT OF THE INTERNET AND THE WEB

The Internet and the World Wide Web have enabled a global information and communications revolution. Firstly, the Internet has triggered the pervasiveness of instant communications. Today it is possible to instantly locate and talk to people, via textual chats or online multicast videoconferencing systems. We are able to perform telephone calls from software to landlines using different technologies such as SIP phones, Skype, or WebRTC for web-based videoconferences. Prices are now much cheaper than the traditional high prices of calls from fixed lines to fixed lines. Email technology has become a vital tool for any organization or person. We can send messages and data over the Internet to any person in the world using a cell phone or laptop when connected to the Internet with just one touchscreen tap or mouse click.

Similarly, the Web provides many valuable services which have made our lives easier, including allowing anyone to publish and discuss his or her interests through services such as blogs, discussion forums, podcasts (personalized on-demand audio and video shows), etc. While the Web began life as a research application, it has rapidly become mature and has been adopted at a faster rate than any other communications technology thus known.

In addition to organizations, anyone is now able to have a personal website (usually for free). Blogs have also become quite popular over the past 5–10 years: these are systems whereby a person can self-publish any content he or she likes. In the same manner, other types of blogs and channels enriched with audio or video are starting to gain in popularity (e.g., SoundCloud, Vine, etc.). On Vine, the content is distributed as a short video clip produced by any user of the service (often armed with just a mobile phone camera). More recently, we have been witness to the YouTube phenomenon, one of the most visited sites on the Web and a major source of all online video traffic.

Over the past two decades, the Web has become one of the most popular transmission platforms for various forms of entertainment, and in parallel it has developed into a recognized global ecommerce framework. It is a new sales channel, a new learning space, a new communications metaphor, and so much more. The Web has reached almost every known organization and today it is possible to find online counterparts to many established institutions: universities, schools, banks, travel agencies, cinemas, etc.

Many different ways and paradigms of carrying out ecommerce on the Web have also been established, for example:

- Business to Consumer (B2C) where an organization's site is focused on selling products or services to the end consumer, e.g., Amazon.com, barnesandnoble.com, etc.

- Business to Business (B2B) where an organization's site is oriented to sell to another organization. A good example of a B2B site is Alibaba.

- Consumer to Consumer (C2C) where any person can sell or buy services from other persons or companies. One of the most popular sites of this type is eBay, and other prominent examples include Japanese site Kakaku.com and Etsy.

- Mobile to Consumer (M2C) where an organization has developed a mobile or responsive website that has been enhanced for mobile devices smartphones or tablets.

These new paradigms have brought about many advantages because—via the Web—it is now possible to buy almost any product from a variety of countries at a lower cost than was previously possible, creating new opportunities for both companies and individuals.

1.4 WEB MINING, THE SOCIAL WEB, AND THE SEMANTIC WEB

As the Web grows larger (in terms of documents and servers serving those documents) and ever more diverse (in terms of content types, topics, languages, etc.), it becomes more difficult for users to find relevant information or to carry out tasks as efficiently. Therefore some mining of relevant knowledge and patterns can assist users in navigating through this ever-expanding information resource. We will discuss various "web mining" techniques in Chapter 2.

The Web was originally envisaged to be not just an information resource, but a Web where users could collaborate and communicate with others around shared topics of interest, and also one where machines could interpret the data that the Web contained to help users with their daily online activities.

As mentioned, the Web has evolved in the first direction to have a more social aspect whereby users can interact with each other, often around content items like videos or status updates. We refer to this in the book as the Social Web (previously referred to as Web 2.0; O'Reilly [2005]), and incorporates aspects such as social software and social media. We discuss the Social Web in more detail in Chapter 3.

In parallel, various efforts have been ongoing in the second direction to make portions of the Web more machine readable for various reasons: data reuse, interoperability and interpretability, improved search around knowledge objects, etc. This direction is called the Semantic Web, and we will give an introduction to its underlying framework in Chapter 4.

1.5   CONCLUSION

In this chapter, we have introduced the book and given a short history of the Web. In the next chapter, we will look at web mining processes and frameworks that can be used to cluster and derive semantic information from the Web (and the Social Web).

CHAPTER 2

Web Mining

In order to give the reader some context so as to understand the following sections of the book, we will describe how the area of web mining was born. Afterward, we will explain in detail how to perform data mining on any type of website, how data mining approaches have evolved to answer a variety of questions, and how they can fulfill a website maintainer's information needs. We will explain in detail the differences between online enhancements and offline enhancements, along with the general process for web mining.

Finally, we will describe some of the clustering and classification algorithms most used in web mining. We will also discuss their relative benefits and disadvantages.

2.1 MINING THE WORLD WIDE WEB

The Web was created through the combination of the Internet and HTML as mentioned in Chapter 1. This very simple idea has allowed anyone to publish their own worldwide-viewable documents without any external control or editorial supervision. In the beginning, much of the content on the Web was produced by researchers, while later, the simplicity of HTML and the complete freedom to publish anything generated a large problem [Pal et al., 2002]: how can one find useful information in this sea of unlabeled, semi-structured, distributed, heterogeneous, high-dimensional, and time-varying documents? The solution was the application of data mining techniques to web-originating data, and in this way, the field of "web mining" was born.

To better understand the origins of *web mining*, we need to go back about three or four decades when searching for relevant content in a huge amount of text documents was impossible for a human being. *Information Retrieval* (IR) systems arose as a solution that helped people to find the information they needed using a computer.

Most IR systems function through the use of *keywords* [Turney, 1993b, 2000, 2003] that the user must provide. Afterward, the system will look for those keywords and retrieve a ranked list of documents in which they were found. This ranking represents the degree of importance or relevance that a results registry has with respect to the query, and usually is represented by a percentage (0% being not important and 100% very important). Most of these rankings are obtained using some calculation based on the number of keywords hits (see Baeza-Yates and Ribeiro-Neto [1999], Schocken and Hummel [1993] for a detailed description). Many of these systems were based on the *Vector Space Model* (VSM) [Saltón et al., 1975] developed by Salton et al. in 1975. This model is still used, and also for the offline website enhancements that we will explain in a later section.

One of the first search engine applications developed on the Internet was Archie: this software worked by periodically searching for openly available and known FTP sites, retrieving a list of their files, and building a searchable index. The commands to search Archie were UNIX commands, and therefore required some knowledge of UNIX to use it to its full capability.[1]

As the Web began and continued its growth, it became necessary to have search engines that would allow people to locate and browse web documents. Because of the huge amount of documents available, initially IR-type techniques were applied. However, the more that the Web grew, the more that IR techniques became useless. This was because the Web was an entirely new space that was different from a traditional set of "research" documents where an editorial analysis is performed and documents have a title, abstract, keywords, etc.

Usually webpages do not have such a formal structure, and can contain many images or multimedia files which may contain the majority of the information of the document. A webpage also may not be a single document: on the contrary, what the reader usually sees is a composition of blocks, dynamically generated from some database backend or maybe even an external source/widget. For these reasons (and many others), it has always been difficult to create efficient search engines for the Web.

The necessity to successfully build server-side and client-side technology in order to better browse across web documents became the main focus of researchers from areas such as information retrieval, knowledge discovery (KD), machine learning (ML), artificial intelligence (AI), amongst others [Pal et al., 2002]. This gave birth to the field of Web Mining (WM), which applies techniques from all of these fields in order to uncover more knowledge from the Web.

Pal et al. (in Pal et al. [2002]) also mention the application of soft computing to web mining, which they called *soft web mining*. Soft computing is a name sometimes given to a set of techniques such as ML, fuzzy sets, genetic algorithms (GA), artificial neural networks (ANN), etc. In essence, web mining is the application of these techniques to web data, so the term *soft web mining* is somewhat redundant and can just be called web mining.

2.1.1   DIFFERENT TYPES OF WEB MINING PROCESSES

A challenging question for website maintainers and organizations often arises: how does one develop a "good" website? In other words, in order to build a website, how does one know that they are giving valuable information to visitors? Is it possible to create a visual design that will allow simple access to the information in a website without confusing visitors? And how can one reliably provide the information, products or services that visitors need? These questions have no simple answer.

Nielsen Nielsen [1999] discusses several website usability problems and also establishes that a website's effectiveness is strongly related with its usability. In addition to the web interface, it is possible to improve the visitors' browsing experience by enhancing the sites' structure and content, but again, how does one do so? This is also a complex question.

---

[1]http://www.walthowe.com/navnet/history.html (updated September 2012, accessed December 2014).

In the late 1990s, several sub-classes of web mining (and associated research lines) were created to help improve the web experience for website users and maintainers, where these depended on the kind of data to be analyzed or enhanced and the type of results that one wanted to obtain and show to a user or maintainer. We will now describe some of these types of web mining processes:

- *Web Text Mining (WTM):* This is the application of the aforementioned techniques (GA, ML, ANN, etc.)—which we will call "mining techniques" from now—to the free text in web documents. After the application of such text processing techniques, we obtain (potentially interesting) text patterns, and eventually deduce some conclusions about how to perform smart improvements on the sites' content, e.g., to extract keywords for document indexing that might be used in a (local) search engine. Some examples can be seen in Bestgen [2006], Chakrabarti [2002], Richardson et al. [2006], Ríos et al. [2005a], Soderland [1999], Turney [1993a, 2001, 2002], Turney and Littman [2005].

- *Web Content Mining (WCM):* This area aims to apply mining techniques to all contents of or objects in a web document. This implies that not only the text is analyzed, but also whole paragraphs, images, videos, sounds, etc. In this sense, WCM is a generalization of WTM. The knowledge extracted could be used in a search engine, but also one is able to find content patterns which can be used to improve the content of a website [Chau and Yeh, 2004, Chen and Chue, 2005, Jin et al., 2005, Liu and Chen-Chuan-Chang, 2004, Norguet et al., 2006b].

- *Web Structure Mining (WSM):* This is the application of mining techniques to the link structures between the pages of a website, and solves a problem with WTM/WCM whereby the visitors' opinion is not considered, i.e., "what is interesting to a visitor" was not included in the analysis. The main idea is to find patterns in the browsed pages to improve the website structure. The results of WSM are usually used to reduce the number of clicks the visitor/user must perform in order to reach what they need. It is also used when pages that are being accessed frequently are placed very deep in the website structure, so as to let website maintainers correct this situation. Some good examples can be found in Berendt and Spiliopoulou [2001], Gery and Haddad [2003], Spiliopoulo [1999], Spiliopoulou [2000], Spiliopoulou et al. [2003, 1999], Srivastava et al. [2000].

- *Web Usage Mining (WUM):* This technique is very similar to WSM, however, the main goal is focused on studying how visitors use a website rather than understanding the structure of the website, using browsing behaviors stored in the logs of the web server. The most famous and common application of this type of mining technique is the generation of browsing recommendations, i.e., while a person is browsing a website, the site predicts the next pages the visitor may like to see and presents this as a suggestion for continued browsing of the site. Some examples are shown in Cooley et al. [1999], Jin et al. [2004], Mobasher [1999],

Norguet et al. [2006a], Pitkow [1997], Rí os et al. [2006]. Simple techniques to analyze browsing sequences were initially developed, and later, complex ones that analyzed visitors' browsing paths [Berendt and Spiliopoulou, 2001, Mobasher, 1999, Spiliopoulo, 1999] were added. Furthermore, some researchers Ríos et al. [2005b] have developed techniques which included the textual content and the sequence analysis to give results closer to visitors' preferences.

- *Web Personalization or Web Profiling (WP):* The aim of this technique is the classification of users in several predefined or dynamically generated profiles. In this way, a site can show contents or offer products/services which a person of a specific profile is more willing to like and use, and therefore buy. This type of technique is commonly used in "profile-based adaptive websites" which apply this technique to registered user accounts. More complex versions of the same technique are used for the same sites but where the user has not registered or logged in. More detailed information is available in Eirinaki et al. [2004], Eirinaki and Vazirgiannis [2003], Eirinaki et al. [2003], Etzioni [1996], Mobasher [1999], Mobasher et al. [2000, 2001], Mulvenna et al. [2000], Perkowitz [2001].

Although all of the above techniques have been shown to be real and effective aids for website content and structure enhancements, they miss the semantic relations among web documents that could be used in the analysis. Some researchers are working on the development of techniques to solve this semantic gap. One way to do so is through the use of an ontology for performing a semantic analysis. However, the development costs of a domain ontology for a small-to-medium sized website can rarely be justified.

Standard processes (e.g., as used in WUM) omit the semantic information from web documents and may lead to poor, erroneous, and hard-to-interpret results that are different from true visitor preferences. This is why a new class of mining techniques, which includes the semantic relations from these web contents, arose. These techniques are referred to as *semantic web mining*.

When working with smaller websites, usually the website experts have a good feeling for what their visitors may want. However, most tools are focused on the automatic discovery of knowledge without considering an expert's previous knowledge of the website or its visitors in the mining process. This produces a huge number of patterns which must be interpreted by an analyst in the case of offline website enhancements. Sometimes, the analysis and understanding of the patterns becomes a very complex and subjective problem.

2.1.2   DIRECTED VS. UNDIRECTED WEB MINING PROCESSES

Based on Berry and Linoff's books, Berry and Linoff [1999, 2000], we will mention two different styles of web mining. *Direct web mining* is used when we know what we are looking for, and usually is implemented as a predictive model where we know exactly what we want to predict. Such is the case for online webpage recommendations, where we would like to predict the next page that a visitor sees. *Undirected web mining* lets the data speak for itself and lets the analyst/expert decide whether the patterns are important/useful or not. This last method is the one used for offline website enhancements, since the analyst does not know if content, structure, or both types of enhancements are required. This way, we let the data show some patterns, which the analyst must interpret to perform any offline modification of the website. This is a very important point: undirected web mining is by necessity an interactive process, where the person who can say if the patterns are useful or not is the topic expert.

2.1.3   SUPERVISED VS. UNSUPERVISED ALGORITHMS

Directed and undirected web mining styles are implemented by using different kinds of algorithms, such as predictive algorithms, clustering algorithms, classification algorithms, etc. In the case of offline website enhancements we aim to obtain knowledge about how the visitors browse the website and what contents are interesting to them. Therefore, most solutions use clustering algorithms over usage data (like web logs) and content information (like the text inside webpages). In order to apply clustering algorithms, we first need to decide which type of algorithm is suitable for our data collection.

Clustering algorithms are divided into two types: supervised and unsupervised. Supervised ones are designed to be applied over already known class labels [Theodoris and Koutroumbas, 2003]. However, sometimes it is not possible to rely on having data already labeled into known classes. Therefore, some algorithms that put this data into one or more classes have been developed. These are called unsupervised pattern recognition or classification algorithms.

In the case of unsupervised algorithms, the problem is to discover the underlying similarities in a set of feature vectors, which represent the data, and then to group these vectors into clusters. For this reason, unsupervised algorithms also are called *clustering algorithms*.

In web mining, it is very difficult to obtain labeled web documents. Therefore, it is a common practice to use unsupervised algorithms for web mining and also for the offline website enhancements problem.

2.2   TRADITIONAL FRAMEWORK FOR OFFLINE WEBSITE ENHANCEMENTS

Most research [Berendt et al., 2002, Berendt and Spiliopoulou, 2001, Chakrabarti, 2002, Etzioni, 1996, Pal et al., 2002, Ríos et al., 2005a] on web mining says that to perform any knowledge mining technique, four subprocesses are needed as shown in Fig. 2.1. However, in the case of offline website enhancements, we need to add an additional fifth stage or subprocess, which is the incorporation of the enhancements themselves into the website. Therefore, overall the subprocesses consist of:

1. Resource Discovery/Data Selection: we select the data sources, number of site versions, time periods, etc.

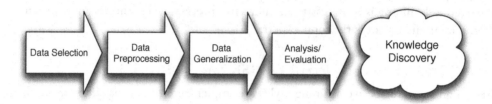

Figure 2.1: Framework for knowledge discovery in databases.

2. Extraction/Data Preprocessing: cleaning the data is a very important task in order to obtain good results. For web text cleaning, a stopwords list, stemming algorithms, spell checking, etc., are used. Also, errors need to be cleaned from the web logs (errors such as 404, 403, etc.). In addition, every time the visitor requests a page, the embedded content (images, sounds, videos) is requested and request times are added into the web logs (which may not be needed). Afterward, a sessionization process over clean logs is performed to rebuild visitors' sessions.

3. Data Generalization: this subprocess consists of the application of one (or more) classification techniques to obtain patterns. For example, a self-organizing feature map (SOFM), OPTICS, or DBSCAN could be used, amongst other techniques.

4. Analysis/Evaluation: we must establish if the resulting patterns are useful or not, and if the classification or clustering is correct or not. Usually this process is performed with the website expert [Berry and Linoff, 2000].

5. Website Enhancements: The analyst decides what will be the specific modifications that must be introduced into the website, both content modifications and structural modifications. Also, we need to establish how and when the enhancements will be introduced.

The offline website enhancement framework as shown in Fig. 2.2 is based on the Knowledge Discovery in Databases (KDD) framework which is shown in Fig. 2.1. As the reader will see, they are almost the same. In this case of offline website enhancements, we show a more detailed solution in Fig. 2.3 which makes use of the basic offline enhancements framework from Fig. 2.2.

The solution presented in Fig. 2.3 uses two basic sources of information: webpage texts and/or contents (depending on the site type as explained previously) and the page requests as stored in the web logs. Afterward, the information must be preprocessed and cleaned to produce good quality results. Next, the information is mixed up and input into a clustering algorithm

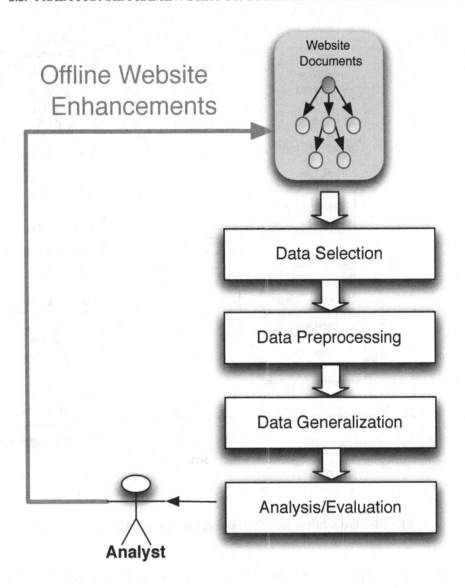

Figure 2.2: Framework for offline website enhancements.

which will return several patterns. The results of this clustering phase must be evaluated and analyzed by the expert who decides which patterns are useful. Based on these patterns, the expert finally decides to perform content and structural enhancements to the website.

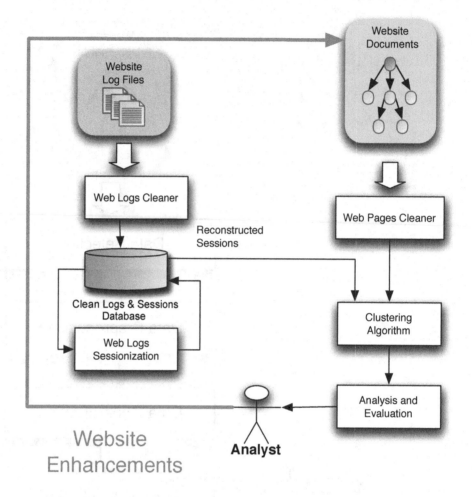

Figure 2.3: More detailed solution for offline website enhancements.

In the following subsections, we will explain each step of the solution presented in Fig. 2.2 and Fig. 2.3.

2.2.1 RESOURCE DISCOVERY/DATA SELECTION

This stage is focused on the automatic retrieval and storage of documents relevant for the analysis while trying not to include irrelevant ones. Typically this process is based on IR techniques [Baeza-Yates and Ribeiro-Neto, 1999, Kosala and Blockeel, 2000, Pal et al., 2002] which include indexing and searching. Usually we need to select the data that is going to be analyzed, then identify the relevant and irrelevant data. Once this is decided, the process to download the data and store it in a suitable way is next. This process is usually performed by crawlers, spiders, robots, etc., which are needed in order to search the Web and download the relevant pages.

When storing the documents, we typically index them. To index a document we need to generate a list of terms which point to specific phrases or paragraphs in the document that contain the information relevant to those terms. Most indexing systems use a database: some of these take a snapshot of the website and then store the resulting index in a database. Four approaches exist for the indexing process Pal et al. [2002]: human or manual indexing, automatic indexing, intelligent or agent-based indexing, and metadata-based indexing.

These techniques are often used in search engines like Google or Bing who have developed huge indexes of big portions of the Web using robots or crawlers. Another example of these methods are metasearch engines, which aggregate and organize the Web's content from several of the leading search engines. A good example is Mamma,[2] which was founded in 1996 and is one of the oldest metasearch engines on the Web.

2.2.2   EXTRACTION/PREPROCESSING

After the relevant information has been stored and indexed (if necessary), a more complicated task arises: How can we automatically extract relevant content from webpages without human intervention: The most common tools we have to do so are the well-known *wrappers* [Muslea et al., 1998, 2001]. The main idea is to use different websites as sources of knowledge. Afterward, information is extracted from pages from these websites using systems that extract relevant text fragments based on a library of many different wrappers, each one used for a specific task. We can say that every wrapper is an *information extraction* (IE) system [Kushmerick, 1999]. An alternative method for IE is proposed in Freitag [1998], where each document is summarized by a set of questions that can be solved by the contents of that document. Therefore, the problem of IE is reduced to only finding the text fragments which answer those questions.

Although IE systems can be very useful, it is not feasible to build scalable systems to process the Web (due to its size and continuous changes). Most systems are developed for particular websites, and some of the most famous systems developed for this purpose include Harvest [Bowman et al., 1995], Faq-finder [Hammond et al., 1995], Information Manifold [Kirk et al., 1995], OCCAM [Kwok and Weld, 1996] and Parasite [Spertus, 1997].

Web text cleaning and preprocessing: stopwords and stemming
As mentioned in Section 2.1.1, all of the different mining techniques (WTM, WCM, WSM, etc.) are based on the same general process as shown in Fig. 2.1. However, every mining process will have its subtle differences in the implementation of each of the four subprocesses.

In order to obtain the best possible results when performing WTM, a website with a relatively high amount of text is needed, with few images, videos, audio, embedded Flash media

---

[2]http://mamma.com/.

objects, etc. This is because on the Web, multimedia files are quite often not labeled by their authors, and we are still some time away from having software that can automatically label all images or videos. However, we can use the HTML field called *alt* for mining some information from these types of content. By doing so, we are now in the WCM area.

After selecting a website that fulfills the above requirements, it is very common to filter out all of the non-useful words in order to reduce the dimension of the word vectors. Afterward, we simply apply the clustering algorithm to the most relevant words (for example, the prepositions, conjunctions, and articles are omitted). Most authors focus on the nouns, adjectives, and verbs. In this way, we can obtain better results and performance improvements. Such a task is commonly carried out through the application of stopword lists.

An interesting problem appears with respect to the plural of words. Should two different words, for instance "car" and "cars," which represent the same concept, be considered as only one term? Another similar problem is verbal conjugations, for instance "drive," "drives," "drove," "driving," etc. In other words [Chakrabarti, 2002], *how do we deal with morphological variants of the corpus?*

This can be solved using *stemming* or *conflation* algorithms, which aim to reduce a word to its root. For example, the word "car" and "cars" are reduced to the same root word "car." Another example are the forms of the verb "drive," which are all reduced to the imperative form "drive." In this way, it is possible to compare two variations of a word.

A commonly used conflation algorithm is the one developed by Porter in Porter [1980]. This algorithm has been modified to process many "Latin" languages such as Spanish, Italian, French, Portuguese, etc. A similar technique can be performed through a dictionary lookup, for example, using WordNet [Miller et al., 1993]. We can take every word and search in a dictionary or thesaurus in order to establish its basic form or word root.

As an example of the application of these techniques in our experiments, after applying Porter's algorithm to a selected version of a website, we reduced the universe of different words to approximately 64%. This allows us to perform the next steps faster and obtain better results.

The main drawback of stemming techniques occurs when two words, let's say W_1 and W_2, with different meanings are reduced to the same root, producing erroneous results when comparing or searching for similar documents. For example, searching for W_1 will return documents which contain this word but will also return documents which contain W_2.

Vectorial Representation of Web Documents

Continuing with the pre-processing problem after cleaning the textual information, we need a suitable representation of documents in order to apply pattern classification/clustering techniques or machine learning techniques, as was partially solved by Salton in 1975 [Saltón et al., 1975]. He proposed a Vector Space Model (VSM) to represent the documents as a feature vector, and in this way, we are able to represent the features as axes in the Cartesian plane. Therefore, every feature vector is represented in the Cartesian plane. An example is shown in Fig. 2.4, where $V1$

and $V2$ are represented in the 3D plane. Both vectors should have the same dimension, so in this case we observe the three dimensions.

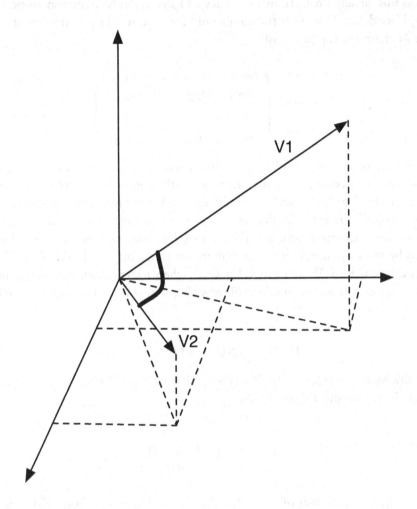

Figure 2.4: A VSM used to represent webpages.

Let W be the number of different words in the entire collection of documents and Q be the number of documents. In our case, a document would be a webpage and the collection of documents the respective website. A vectorial representation of the website would then be a matrix M of dimension $W \times Q$ with:

$$M = (m_{ij}) \ \forall i = 1, \ldots, W \ \wedge \ j = 1, \ldots, Q \tag{2.1}$$

In Eq. (2.2), the variable m_{ij} is the weight of word i in document j. This weight must capture the fact that a given word can be more important than another one. Many slightly different ways to do this already exist. Term Frequency (TF) is the most basic way to perform a term's weighing. If we define $TF(i, j)$ as the number of times that word i is in document j then we can fulfill the M matrix in Eq. (2.2) easily.

$$M = \begin{pmatrix} m_{1,1} & m_{1,2} & \cdots & m_{1,Q} \\ m_{2,1} & m_{2,2} & \cdots & m_{2,Q} \\ \vdots & \vdots & \vdots & \vdots \\ m_{W,1} & m_{W,2} & \cdots & m_{W,Q} \end{pmatrix} \quad (2.2)$$

However, we face the problem that some words which are repeated in many documents of our collection can become very important while they are not, i.e., "the," "a," "an," "and," or words like "student" or "professor" which are repeated many times in different documents of a university website. To reduce the effect of these words in the entire collection, we need some way to attenuate them. Inverse Document Frequency (IDF) may be used to reduce the noise effect introduced by those unimportant words that are too frequently used in different documents. A common expression for IDF is shown in Eq. (2.3), where n_i represents the number of documents where $term_i$ appears. It is very simple to observe that if n_i is very close to Q then $IDF(i)$ will be very near to 1.

$$IDF(i) = \lg(Q/n_i) \; \forall i = 1, \ldots, W \quad (2.3)$$

Finally, we combine IDF with TF in the traditional way in Eq. (2.4). This expression allows us to obtain better weights for the words.

$$\begin{aligned} m_{ij} &= TF(i, j) * IDF(i) & (2.4) \\ &= f_{ij} * \lg(Q/n_i) & (2.5) \end{aligned}$$

where f_{ij} is the number of times that the ith word in the jth page and n_i is the number of documents containing the ith word. A page p_j is represented by the column j in M, i.e., $p_j \rightarrow (m_{1j}, \ldots, m_{Wj})$ see Eq. (2.1) and Eq. (2.2). Eq. (2.4) is widely used in the area of "web intelligence."

Another method to give weights to words is by considering some as "special words." This system works by defining words in the title of a webpage, those that are underlined, in italics, or bolded as special words. These words are usually marked thus because the author wants to emphasize something, which also means that a greater weight should be given. Therefore, a $TF *$ IDF-based method can be combined with the special words. This is carried out through the use of a factor α that increases the weight of a word if it is a special word (and do nothing if it is not).

2.2.3   DATA GENERALIZATION

This task is where we apply techniques from the pattern recognition area or the machine learning area to extract information from web data. Pal et al. [2002] affirm that most machine learning systems learn more about a user's interests than the Web itself.

Depending on the selected technique for pattern recognition or learning, many problems arise. One of them is that many techniques require some positive examples and negative examples to be classified beforehand. However to do so, some human intervention is needed, and this is almost impossible to carry out for large portions of the Web or for complex patterns.

Association rule mining is also part of this phase. An association rule is an expression like $X \Rightarrow Y$ where X and Y are rules, and a rule is a set of items. For example, let us say we have a transaction T_1 that contains the following items $T = \{a_1, a_2, a_3, a_4, a_5\}$, and we have the rules $X = \{a_1, a_2, a_3\}$ and $Y = \{a_4, a_5\}$. Through the association rule $X \Rightarrow Y$ for some transaction which is recommending the items in X to a user, the system will also recommend the items in Y to that user. This is a very simple example and we are ignoring the probabilities involved in the process.

2.2.4   ANALYSIS/EVALUATION

The word analysis implies reasoning, which only humans are able to perform when there is no previous knowledge. For this reason, the evaluation process is a crucial task, which all web miners should perform carefully. It is also important to notice that humans play an extremely important role in knowledge extraction, since machines are still incapable of human reasoning. Therefore, humans should always supervise the validation and interpretation of discovered patterns.

Once the generalization phase finishes, the analyst can use several tools in order to visualize the information. This can be through tools like WEBMINER [Mobasher et al., 1997], Webviz [Pitkow, 1997], or by looking at the work presented by Cohen in Cohen [1995] where the author focuses on the nature of the knowledge that can be extracted from the Web.

We must emphasize that the usefulness of results must be verified by asking a group of end beneficiaries. This is because we might discover some patterns which are useful and some which are not, but this primarily depends on the domain owner or final users.

2.3   CLUSTERING ALGORITHMS FOR WEB MINING

We are now going to explain several different types of algorithms, which are used in (offline) website mining. Since this area is quite large, we are going to focus on the most important algorithms used to perform the generalization subprocess. We have in mind that we are working with an undirected approach (Section 2.1.2) and because of this we will choose an unsupervised algorithm (Section 2.1.3).

The clustering problem consists of creating groups of objects with similar characteristics [Berry and Linoff, 1999, 2000, Duda et al., 2001, Himberg et al., 2001]. In the case of clustering, we do not know the labels of the objects beforehand, which makes it different from the classification problem.

Let Ω be a set of m vectors $\omega_i \in \Re^n$, with $i = 1, \ldots, m$. The clustering goal is to partition Ω in K groups, where C_k denotes the kth cluster. Then, $\omega_i \in C_l$ means that ω_i is more similar to the elements in this cluster than to objects from other clusters. The criteria to establish if an element is similar to another is a function called the similarity measure.

2.3.1   CLUSTERING METHODS

We could list many different methods to solve the clustering problem, however, in general we will highlight three:

Partitioning Methods: The idea of these methods is to divide n data vectors into k groups/clusters called partitions. Let $X = \{x_1, \ldots, x_n\}$ be the dataset and $P = \{p_1, \ldots, p_k\}$ be the partitions desired, with $k \leq n$. A partition p_i is a subset of X such that:

- $p_i \neq \phi$.
- $X = \cup_{i=1}^{k} p_i$.
- $p_i \cap p_j = \phi$ $\forall i \neq j$

Hierarchical Methods: The main idea is to construct a hierarchical decomposition in a set of data using either an agglomerative or divisive method. Using a hierarchical agglomerative approach, we create as many groups as data vectors and then we place each data vector into one group. Next, similar clusters are merged until the desired number of clusters is obtained or another ending condition is reached. The hierarchical divisive method places all data vectors into one cluster. Afterward, this cluster is divided successively into smaller ones until the terminal condition is reached.

Density Based Methods: These methods are based on the concept of the density of particles as used in physics. Clusters are discovered when a neighborhood exceeds a density threshold. Let $X = \{x_1, \ldots, x_n\}$ be the dataset, $C = \{c_1^t, \ldots, c_m^t\}$ the clusters set in the iteration "t", δ the density threshold and $card(c_k)$ the number of elements in cluster c_k. A data vector x_j belongs to c_k when $D(c_k^t, x_j) \leq R$, with D a distance-like function and R a radius. The process stops when $\forall c_k \in C, card(c_k) \leq \delta$, otherwise the cluster centroid is redefined.

2.3.2   SELF-ORGANIZING FEATURE MAPS

This algorithm, also called SOFM or simply SOM, is commonly used in unsupervised learning for data mining [Berry and Linoff, 1999, 2000, Chakrabarti, 2002]. It represents the result of a vector quantization process. The SOFM maps the input space into a bi-dimensional array of nodes that are also called neurons. The array's lattice can be rectangular or hexagonal. Every neuron is an array of similar dimensionality to data vectors. Let us call $m_i \in \Re^n$ the neuron i in an SOFM, and the components of every neuron are called the *synaptic weights*.

To begin the SOFM algorithm, all neurons must be initialized. This process is performed by creating random synaptic weights for every neuron. Afterward, we commence the learning or training phase of the SOFM. Let $x_i \in \Re^n$ be an input data vector, we present x_i to the network, and using a metric (similarity or distortion measures), we determine the most similar neuron (center of excitation, winner neuron, best matching unit (BMU), or centroid). This process is performed for every input example x. Once all data vectors have been presented to the network, we say an *epoch* has finished. Next, we must begin another *epoch*: this is done by presenting all data vectors to the network again. Finally, after many epochs, we obtain convergence of the SOFM and we can finish the training phase.

The *best matching unit* (BMU) is obtained from Eq. (2.6), and if the SOFM has N neurons, then $m_c(t)$ is the winner neuron, defined as the BMU in the whole network for the example x_i in the epoch t.

$$||x_i - m_c(t)|| = \min\{||x_i - m_j(t)||\} \quad or \quad c = arg\min\{||x_i - m_j(t)||\} \quad \forall \quad j = 1, \ldots, N \tag{2.6}$$

Once the winner neuron (BMU) has been identified, we activate it to actually learn the example vector. To do so, we use a learning function as shown in Eq. (2.7). An important factor is that the learning function alters the synaptic weights of the BMU, but it also alters the weights of the surrounding neurons to a lesser degree. In this way, the BMU is moved toward the data example, and the surrounding neurons are also moved, but moved less than the BMU. This effect depends on the distance from the BMU (or centroid) as shown in Eq. (2.7) and is transmitted to all neurons in the network.

$$m_j(t + 1) = m_j(t) + h_{cj}(t) \cdot (x_i(t) - m_j(t)) \quad \forall \quad j = 1, \ldots, N \tag{2.7}$$

Eq. (2.7) shows the basic weight modification algorithm, where t is an integer and represents the iteration, and h_{cj} is called the "neighborhood kernel." It is a function defined over the lattice points, and usually $h_{cj} = h(||r_c - r_j||, t)$, where $r_c, r_j \in \Re$ is the radius between the BMU and another neuron in the array. The expression h_{cj} is such that when $||r_c - r_j||$ is increased, $h_{ci} \to 0$.

Depending on the points chosen by h_{ci}, it is possible to define different notions of neighborhood. Therefore, we are able to define diverse topologies of the SOFM (see Fig. 2.5), such as:

• Open topology: The idea is to maintain the bi-dimensional space geometry when carrying out the learning. The edges of the map are not connected to other neurons, producing an abrupt end to the learning. This effect is more notorious when the BMU is closer to the edges of the map.

- Tape/cylindrical topology: Using this topology the idea is to connect two borders of the bi-dimensional map. Thus, the learning effect continues to the opposite side of the map, although the other edges of the map are disconnected.

- Toroidal topology: In this topology, we connect all borders of the bi-dimensional grid. In this way, we never finish the learning effect in any direction. We always transmit the effect to all neurons in the map in a smooth way, and we say that this topology helps to maintain the continuity of the space.

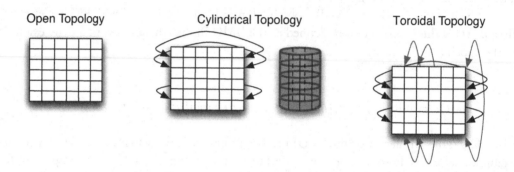

Figure 2.5: Possible topologies for an SOFM.

2.3.3   K-MEANS CLUSTERING ALGORITHM

The main idea of this classification algorithm is to assign each vector to a set of given clusters. Clusters' centroids are updated on each iteration using a previously established rule for this purpose. This procedure is repeated iteratively until we reach a stopping criterion. This algorithm falls into the category of partitioning algorithms which places every data vector into one cluster (see Section 2.3.1).

We show an example in Fig. 2.6, where we have three different types of objects. The algorithm starts by setting the number of groups to discover, in this case, three (in real cases, the value of k may be unknown). Then, three randomly created centroids are created, which we call the seeds. The seeds then start to move to the center of the groups by computing the mean among the data vectors, hence the name "k-means." The move toward the center of each group is performed gradually after several iterations.

Instead of defining random seeds, we can randomly select k vectors from our data as seeds. Afterward, the process is performed similarly to that explained above.

The specific steps of the k-means algorithms are as follows:

Given $c_1^t, \ldots, c_k^t$ as cluster centroids on iteration t, we compute $c_1^{t+1}, \ldots, c_k^{t+1}$ as cluster centroids on iteration $t + 1$, according to the following steps:

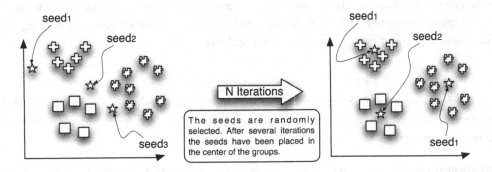

Figure 2.6: K-means algorithm.

1. Cluster assignment: For each vector in the training set, we compute the similarity between this vector and $c_j^t \ \forall j = 1, \ldots, k$ to determine which class it belongs to.

2. Cluster centroid update: Let $V_j^t = \{v_1, \ldots, v_{q_j^t}\}$ be the set of q_j^t vectors associated to centroid c_j^t, with $j = 1, \ldots, k$. The next centroid is determined as c_j^{t+1}, a mean of V_j^t, i.e.,

$$\forall v_i \in \left\{ V_j^t / \max\{ \sum_{t=1}^{q_j^t} sm(v_i, v_t) \} \right\}, i \neq t, \text{ where } sm(,) \text{ is a similarity measure.}$$

3. Stop when $c_j^{t+1} \approx c_j^t$.

Even though this method requires us to know beforehand the number of clusters to be used in the clustering, it is a very commonly used method in web mining due to its simplicity and speed. However, we require several trial-and-error runs to discover the final results. In our case, for offline enhancements based on content and sessions, it is really difficult to know beforehand the number of clusters hidden in the data. For example, in the specific case of web usage mining, there exists as many patterns as browsing page combinations. Therefore, assuming a number of clusters a priori is impossible. To avoid this situation, we often apply other algorithms such as SOFM which may reveal a number of clusters, and then use that to setup the k value with this information.

Another problem with this method when applied to offline enhancements is that all data vectors are placed in one group. In the Fig. 2.6 example, we can clearly observe three classes. However, when working with sessions and textual information, these differences become more subtle and, many times, very subjective. Therefore, establishing if the data vectors are truly members of the class is quite a complex task and the analyst must perform it. In offline enhancements, the number of input data vectors is huge—we are talking about thousands of sessions—therefore if we setup $k = 5$ (as in our experiments), the groups created by k-means can still contain a huge number of data vectors which makes the validation process even harder. Of course, the acquisition of any information to perform an offline enhancement is also very hard, and oftentimes the overwhelming amount of patterns to be analyzed prevents the analyst from discovering valuable information that could be used to improve the website.

2.3.4   DECISION TREES

Decision trees are a very versatile tool used in data mining and web mining. Many different varieties of this algorithm exist, however, we can classify them into two types: classification trees and regression trees.

Classification trees: These trees label records and assign them to a proper group. This type of tree can also give confidence that a classification is correct by showing the probability of the class, which is the confidence that a record belongs to a class.

Regression trees: These trees are used to estimate the values of a target numerical variable. Therefore, we can predict the expected product price that a subscriber may pay for new services.

The different types of trees are based on the same mechanics. A record travels from the base of the tree (also known as the root) and follows a path through several nodes. On every node, we must answer questions like "is field 5 greater than 34?" or "did customer A buy apples, grapes, bananas or oranges?" This is repeated until the record reaches a terminal node or leaf. Then we assign that record to a class associated with that leaf. In the case of regression trees, we can assign a value based on the mean or some other mathematical function for the target variable on that leaf.

In Fig. 2.7, we show a decision tree example used to cluster different objects. In the figure, we only have two variables for simplicity: these are X and Y. X can take values from 1 to 100 and Y from 1 to 200. Then, if we take a record $(X, Y) = (34, 126)$, then the first question in the root node, "is $X \leq 65$?," in our case is "yes." Next, we need to answer a second question, "is $Y \leq 100$?," and the answer is "no." Now, we are in a leaf node, and we can assign this record to the "cross" class.

The application of decision trees for offline website enhancements may look very similar to the example shown in Fig. 2.7 but it is much more complex. Firstly, in order to establish good clustering, we need to know a lot of very specific information about the website visitors. In the case of a university website, we would need information about the age, gender, the ISP of the visitor, etc. These types of information are not contained in web server logs, and many organizations would not have a registration system for their website's visitors. As a consequence, we are not able to apply decision trees to web sessions for offline enhancements. Even if we try to differentiate among the pages visited, the number of conditions is as large as the combinations of pages that the website has (remember that a site may have from hundreds to thousands of pages, and therefore all combinations of these pages would be huge). Consequently, it is also not feasible to apply decision trees to offline enhancements based on web logs and site contents.

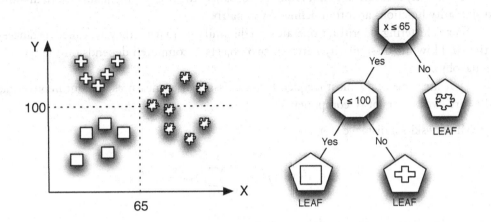

Figure 2.7: A decision tree for clustering.

2.4   DISSIMILARITY AND SIMILARITY MEASURES

The main issue when applying a pattern recognition algorithm is to establish a similarity or dissimilarity measure. Usually in the web mining field, Euclidean distances or cosine measures are used. However, depending on the type of data and the feature representation, it is possible to develop different similarity measures which give better results.

A dissimilarity (or distance) between object x and y (or distance measure) is the function $d(x, y) : X \quad X \to R$ which satisfies the following conditions:

$$d(x, x) = \quad 0 \tag{2.8}$$
$$d(x, y) \geqslant \quad 0 \tag{2.9}$$
$$d(x, y) = \quad d(y, x) \tag{2.10}$$

Also, to have such a distance, we require that the triangle inequality be satisfied, i.e., for any objects x, y and z:

$$d(x, z) \quad d(x, y) + d(y, z) \tag{2.11}$$

A similarity $s(x, y)$ between object x and y is a function $s(x, y) : X \quad X \to R$ which satisfies the following conditions:

$$s(x, x) = \quad 1 \tag{2.12}$$
$$s(x, y) \geqslant \quad 0 \tag{2.13}$$
$$s(x, y) = \quad s(y, x) \tag{2.14}$$

A similarity measure does not require us to satisfy the triangle inequality. Both dissimilarity and similarity functions are often defined by a matrix.

Some clustering algorithms operate on a dissimilarity matrix (they are called distance-space methods). How the dissimilarity between two objects is computed depends on the types of the original objects.

Here are some of the most frequently used dissimilarity measures for continuous data, if x_i and x_j are two vectors of N components.

- Minkowski L_q distance (for $1 \geq q$)

$$d(x_i, x_j) = \sqrt[q]{\sum_{p=1}^{N} |x_{ip} - x_{jp}|^q} \tag{2.15}$$

- City-block (or Manhattan distance or L_1)

$$d(x_i, x_j) = \sum_{p=1}^{N} |x_{ip} - x_{jp}| \tag{2.16}$$

- Euclidean distance (also known as L_2)

$$d(x_i, x_j) = \sqrt{\sum_{p=1}^{N} (x_{ip} - x_{jp})^2} \tag{2.17}$$

- Chebychev distance metric (or maximum or L_∞)

$$d(x_i, x_j) = \max_{p=1...N} (|x_{ip} - x_{jp}|) \tag{2.18}$$

The basic principle used by most similarity measures for web mining is the dot product between vectors. This assumes that our feature vectors are represented as numbers which allow us to represent the features in the Cartesian space, as shown in Fig. 2.4.

In Eq. (2.19), we introduce an expression for the dot product distance PD, also known as a cosine measure, which allows us to compare two web documents' feature vectors p_i and p_j, and where the vectors must be normalized to avoid the problem of scale. The values returned by PD are in the interval $(0, 1)$ as shown in Fig. 2.8.

$$PD(p_i, p_j) = \frac{\sum_{k=1}^{W} m_{ki} m_{kj}}{\sum_{k=1}^{W} m_{ki}^2 \sum_{k=1}^{W} m_{kj}^2} \tag{2.19}$$

Similarity measures may be more complex so as to better represent the differences between objects in complex data types, or to decrease/increase the effect of specific variables from those complex objects.

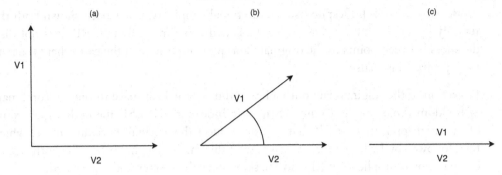

Figure 2.8: The dot product between vectors is commonly used in similarity measures: (a) when the angle between feature vectors V1 and V2 is $\angle = 90°$, the dot product is 0, meaning the vectors are totally different; (b) if the angle is between $(0°, 90°)$, then, the dot product is between $(0, 1)$; (c) when the angle is $\angle = 0°$, the dot product is 1, meaning the vectors are the same.

In Ríos et al. [2006a,b], two types of similarity measures for web usage mining are given, along with their application on real websites for online website reconfiguration.

2.5   LATENT SEMANTICS USING LSA TECHNIQUES

Up until now, we have not talked about the vocabulary size problem, which is an important issue when processing large collections of documents. The amount of features we have encountered in our experiments has been rather high: in our experimental website and online communities, this was between 11,000 and 80,000. To reduce the number of features and produce clusters of higher quality in less processing time, we used stopword lists and stemming (see Section 2.2.2). However, this is a simplistic method to reduce complexity. We refer to four other potential methods [Ampazis and Perantonis, 2004] to reduce the dimension of the histogram vectors, without reducing the quality of the clusters significantly:

1. The projection of the data onto a lower-dimensional orthogonal subspace where most of the variance is concentrated in the new subspace axes. For textual data domains, this method is known as Latent Semantic Indexing (LSI) [Deerwester et al., 1990] and it is based on a Singular Value Decomposition (SVD) of the term-document matrix for the textual collection. In addition to dimensionality reduction, LSI exhibits improved retrieval performance since theoretical and experimental results have shown that it enhances the semantic aspects of the data. A potential problem with LSI arises from its computational cost, since the evaluation of the SVD for high-dimensional datasets can be quite high.

2. Reduction of the dimensionality of the histogram vectors by the multiplication of the term-document matrix with a random matrix (Random Projection (RP) method) [Johnson and

Lindenstrauss, 1984]. Despite its computational simplicity, it has been shown both theoretically [Johnson and Lindenstrauss, 1984] and experimentally that RP does not distort distances between points in the original data space, especially in the case where the matrix to be projected is sparse.

3. Projection of the original data onto a lower-dimensional subspace through a combination of Random Projection and Latent Semantic Indexing (RP/LSI) methods which consists of an initial application of RP, that suitably reduces the original space dimension, which is then followed by LSI. The main advantage of this method is that it benefits both from the computational simplicity of RP and the semantic enhancement of data of LSI.

4. Clustering of words into semantic categories, as is performed in the WEBSOM method [Honkela et al., 1996, 1997, Kaski et al., 1996].

LSA is a technique in natural language processing, in particular in vectorial semantics, patented in 1988 by Scott Deerwester, Susan Dumais, George Furnas, Richard Harshman, Thomas Landauer, Karen Lochbaum, and Lynn Streeter. In the context of its application to information retrieval, it is sometimes referred to as Latent Semantic Indexing (LSI). We use the LSA technique because it gives very good results when indexing large collections of documents as well as showing the semantic relations amongst them, and has become common practice. In particular we use the LSISOM approach given in Ampazis and Perantonis [2004]. This method is used to perform automatic labeling of documents.

LSA uses a term-document matrix, which describes the occurrences of terms in documents. It is a sparse matrix whose rows correspond to documents and whose columns correspond to terms, typically stemmed words that appear in the documents. A typical method for weighting the elements of the matrix is TF-IDF in Eq. (2.4). This is why it is feasible to directly apply LSA to the VSM representation of the website shown in Eq. (2.2) [Bestgen, 2006].

The main idea of LSA is to generate a linear mapping between the original VSM matrix and a reduced dimensional subspace. To do so, LSA methods use the SVD to decompose a matrix into several matrices that accomplish the property shown Eq. (2.20).

Let A be a matrix of frequencies similar to M in Eq. (2.1), and then we can write A as shown in Eq. (2.20).

$$A = U \Sigma V^T \tag{2.20}$$

In Eq. (2.20), matrix U and V are orthogonal and contain the left and right singular vectors of A. The diagonal matrix Σ contains the singular values, which are ordered from higher in position $(1, 1)$ to lower. LSA then achieves dimensionality reduction, maintaining the k-largest singular triplets from this decomposition. It is important to notice that $k < r = rank(A)$. In this way, all the a^j data vectors are projected onto a k-dimensional subspace.

$$\tilde{a}^j = (a^j)^T U_k \Sigma_k^{-1} \tag{2.21}$$

In Eq. (2.20), the dimension of the matrices is $A_{m\ n}$, $U_{m\ r}$, $_{r\ r}$ and $V^T_{r\ n}$. Therefore, $U_{k(m\ k)}$, $_{k(k\ k)}$ and $V^T_{k\ (k\ n)}$. The rows of U_k are considered to be the LSA for the term vectors, and the rows of V_k are considered as the LSA representation of the documents vectors (see Fig. 2.9).

This method is usually applied to large document collections in order to be able to extract semantic information from these patterns [Ampazis and Perantonis, 2004], which is the case when working in web portals, but it is more difficult to apply in websites which only contain hundreds or thousands of pages.

The LSA representation of the term vectors (rows of U_k) identifies similarities on terms, and we are able to apply cosine distances to obtain more insights into the data. The model has the advantage that it suppresses the effect of the variation in the use of a term by assigning similar vectors to the terms used similarly and different vectors to the terms used very differently. When maintaining the first k singular values, we move these terms closer together if they are similar and further apart otherwise [Deerwester et al., 1990].

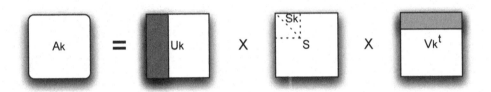

Figure 2.9: Singular value decomposition.

This method has shown relatively good results in representing synonyms. However, it is considered to be poor for representing antonyms and quasi-antonyms [Landauer et al., 1998]. Landauer in Landauer et al. [1998], so as to prevent us from using LSA erroneously, says *"Investigators who use LSA vectors should keep these factors in mind: the similarities should be expected to reflect human similarities only when averaged over many word or passages pairs of a particular type and when compared to averages across a number of people; they will not always give sensible results when applied to the particular words in a particular sentence."* He also mentions that since LSA is a "bag of words" method, it ignores all syntactical, logical, and nonlinguistic pragmatic entailments, resulting in missing meanings or giving scrambled results. This is an undesired effect, since we need to obtain good quality results to allow an expert to extract related information so as to improve a website. In our approach, we avoid this by using experts' criteria in the fuzzification of the terms that represent a concept.

Another useful paper on web text classification was published by Ampazis and Perantonis [2004], where the authors developed LSISOM as an alternative to a WEBSOM. They used the LSISOM approach to process 420 documents with 5,923 different words. These ratios can vary: in our experiment, for one version of a website we had 206 documents and 11,435 different words.

2.6   CONCLUSION

We have given a brief introduction to several text mining techniques that can be used for website mining. These algorithms can be used to extract information from any kind of website ranging from static websites to social websites and networks. However, in the case of mining users' generated data, the task to gather data becomes more complex and it must be developed carefully, since the results of the mining process are strongly dependent on the data quality. In the next chapter, we will look at some of the common characteristics of such social websites.

CHAPTER 3

The Social Web

Now that we have examined the Web and various techniques that can be used to mine more information from documents and activity logs on the Web, we will now move on to examine the first of two parallel and complementary evolutions of the Web: the Social Web.

3.1 WHAT IS THE SOCIAL WEB?

The Social Web is a term that describes one evolution of the Web—from being a place where information pages were created and updated by a few people and displayed to many users—toward a multi-user interactive framework with many people all contributing and sharing content, having conversations around that content, and connecting to like-minded individuals in real-time through a heterogeneous set of interfaces all connected by the Web.

The Social Web has its origins in a variety of Internet technologies that predate the Web, including dial-up bulletin board systems and USENET. The Social Web is also an extension of the term "Web 2.0," and is connected to the concept of social media.

Web 2.0 referred to the next generation of Web-based communities and read-write web systems: to a variety of emerging structures and abstractions that emerged on top of the World Wide Web. A common thread underlying most Web 2.0 systems was that they facilitated collaboration and sharing between users. Web 2.0 was not only a technological evolution but also an emerging business trend, according to Tim O'Reilly [O'Reilly, 2005]: "Web 2.0 is the business revolution in the computer industry caused by the move to the Internet as platform, and an attempt to understand the rules for success on that new platform."

Social media is a more recent term that is often used to refer to web-based platforms like Facebook,[1] Twitter,[2] and YouTube,[3] and the user-generated content that people share on those systems that can be commented on and tagged by others. However, social media too has its origins in earlier types of social software designed for the Web such as discussion boards and guest books. Social media exists as a complementary source of media to traditional platforms like broadcast media, newspapers, etc.

The Social Web has applications on intranets that parallel those on the public Internet. On the Internet, the Social Web enables group participation through the combination of individual user contributions via social networks, blogs, microblogs, and content tagging services, and has

---

[1]http://www.facebook.com/.
[2]http://www.twitter.com/.
[3]http://www.youtube.com/.

unleashed the power of communities with efforts like Wikipedia[4] demonstrating the "wisdom of the crowd" in creating the world's largest encyclopedia. But Social Web technology is also used in company intranets as an effective knowledge management, collaboration, and communication tool.

The Social Web itself is evolving with the computing devices being used to create and view content, as we move from a majority of people using laptops or desktops to access the Social Web, to a majority now accessing the Social Web through smaller devices like tablets and mobile phones. The features of these mobile devices mean that information on the Social Web becomes even more real-time, due to the fact that they are more ubiquitous and often always connected to the Internet.

The Mobile Social Web has resulted in the creation of applications such as Swarm from Foursquare[5] for sharing one's current location, and has also led to a surge in the use of Social Web platforms such as Twitter and Facebook since short updates can be sent more immediately from these systems.

3.1.1   A BRIEF HISTORY OF THE SOCIAL WEB

The first wave of Social Web platforms revolved around web-based discussions, either through Web-only bulletin boards, or through mailing lists and newsgroup discussions that were lifted onto the Web (roughly from 1996 to 2003). This included systems like the Ultimate Bulletin Board,[6] vBulletin,[7] Mailman,[8] and Deja News.[9]

The next wave of Social Web platforms focused on the creation of networks between people, and also incorporated some discussion on functionality, but the primary aim was to create social networks through online platforms (we will refer to these as online social networks or OSNs) (approximately 2004 to 2006) [Boyd and Ellison, 2008]. Early examples of OSNs include Friendster,[10] Orkut,[11] and Tribe.[12]

After 2006, many social networks began to make content the primary objective of using these platforms, and this ties into the notion of object-centered sociality from Knorr-Cetina Knorr-Cetina [1997], whereby people are connected to each other but the stronger networks are formed through common "social objects" of interest. These social objects include videos on YouTube, blog posts on topics of interest, tweets linked to a certain hashtag, and posts on forum-type discussion areas/groups on Facebook, LinkedIn,[13] etc.

---

[4]http://www.wikipedia.org/.
[5]http://www.swarmapp.com/.
[6]http://en.wikipedia.org/wiki/UBB.classic.
[7]http://www.vbulletin.com/.
[8]http://www.gnu.org/software/mailman/.
[9]http://en.wikipedia.org/wiki/Google_Groups#Deja_News.
[10]http://www.friendster.com/.
[11]http://www.orkut.com/.
[12]http://www.tribe.net/.
[13]http://www.linkedin.com/.

3.1.2 ONLINE SOCIAL NETWORKS

The field of social networks has had an inexorable rise in popularity in the last few years. A social network is normally defined as a network of interactions or relationships where the nodes consist of actors, and the edges consist of the relationships or interactions between these actors.

In the early years of the formal study of social networks, structural properties of these networks were much in vogue. The characterization of these social networks was deduced through these structural properties, properties such as homophily (i.e., similar nodes tend to associate together), the small-world phenomenon (i.e., the fact that social networks are so rich in short paths), power-law distributions.

An online social network (OSN) is defined as a social network mediated by the Internet. Online social networks have achieved widespread use and popularity since 2008, with over a billion people using Facebook regularly and hundreds of millions of people using platforms like Twitter and LinkedIn, all of them creating, sharing, and commenting on billions of pieces of content.

OSNs allow a user to create and maintain an online network of close friends or business associates for social and professional reasons. There has been an explosion in the number of online social networking services in the past ten years. But these sites do not usually work together and therefore require you to re-enter your profile and redefine your connections when you register for each new site.

One of the most popular social websites and major contributor to overall web traffic is YouTube, where users upload videos and comment on other people's media. Some also consider YouTube to be a social network of sorts, as users begin to subscribe to channels and follow specific users to be kept up-to-date with their video output.

The rise and rise of online social networks

The last few years have seen the rise in popularity of online social networks such as Facebook, LinkedIn, etc. One of the reasons for their popularity is that they are not constrained by geographical limitations nor by physical presence. These online social networks have a high degree of infrastructure that allows them to run in a reliable fashion, with this infrastructure supporting a rich variety of data analytic applications such as search, text analysis, image analysis, etc.

There are unique issues that arise in the context of the interplay between the structural and data-centric aspects of online social networks. Many of these social networks have several millions of users who post content, such as texts, images, etc., onto these networks. These contents are repositories of valuable information that can be mined for a users' likes, dislikes, tastes, etc., and which can be useful in so many fields of endeavors, for example in sales, defence, finance, etc. Mining these repositories efficiently provides an unprecedented challenge. The effective management of the data, for example, as in storage, is also a challenge due to the very high volume of the available data.

Because of these high volumes of data and valuable information that these repositories contain, another challenge is the task of maintaining the data confidentiality and data integrity of these repositories. Users are finding it difficult to determine the trustworthiness of the information propagating via these social networks.

3.1.3   SOCIAL MEDIA CREATION AND SHARING

Social media content can take many forms on various websites: blog entries, message board posts, videos, audio, images, wiki pages, user profiles, bookmarks, events, etc. Most social websites share a common interaction pattern: a user creates some content item (just the content itself, or the content with a title, description, and other metadata), and then waits for comments and replies from users in their network or perhaps beyond their network.

There is a distinction between content that is created by the user inside the social website, and that which is from outside the social website but is shared into it (i.e., from an external source). For example, a user could choose to write their own report on a sports game and post it on their social media channel for their friends to read. Alternatively, they may be visiting an external website with a news story about the game, and by clicking on a share button (embedded on the site or using an add-on in their browser), the social website would generate a synopsis of that news story in the user's channel (which their friends can then see and comment on).

Beyond commenting, most social websites allow some further interactions between users and content on the social website through re-sharing or favoriting mechanisms. Re-sharing mechanisms include the retweet button on Twitter and the share to profile/group/page buttons on Facebook or LinkedIn. Favoriting mechanisms include the star button on Twitter, the Like buttons on Facebook or LinkedIn, and the +1 button on Google+.[14] (The Facebook Like button has two operations, in that it can be used to mark favorite content, but also to share some content on one's profile page.)

3.1.4   TAGGING, FOLKSONOMIES, AND HASHTAGS

Tagging is common to many social websites—a tag is a keyword that acts like a subject or category for the content it is added to and becomes associated with. Combinations of tags can yield folksonomies: collaboratively generated, open-ended labeling systems.

Folksonomies enable Social Web users to browse and navigate the content that has been tagged, and to visualize popular tag usages via "trending topics" or "tag clouds" (visual depictions of the tags used on a particular website, similar to a weighted list in visual design).

Folksonomies are one step in the direction of the Semantic Web. The Semantic Web often uses top-down controlled vocabularies to describe various domains, but it can also leverage folksonomies and therefore develop more quickly since folksonomies provide a distributed classification system with low entry costs.

---

[14]http://plus.google.com/.

A hybrid tagging-classification system that has been leveraged by the Semantic Web is the Wikipedia categorization system. Any user can tag a Wikipedia article with a keyword denoting a category, but these categories can also be linked to each other. Because of the broad scope of Wikipedia, this has created a classification or category hierarchy that is now being used to group semantic entities and link different semantic datasets to each other.

Hashtags are a variation on tags that were originally used on Twitter by Chris Messina.[15] With the form #keyword, users often use these tags at the end of a microblog post, or sometimes within the text of the microblog post itself (e.g., "I went to see the new #Lego film today #awesome"). Hashtags can also be used to group microblog posts related to a particular real-time event, e.g., a sports match, TV show, breaking news story, or product launch. They have since been adopted by Facebook.

3.1.5 CROWDSOURCING AND CITIZEN SENSORS

Through a multitude of social media contributions from a diverse set of people across the globe or in a specific geographic location, we can combine multiple knowledge facts and opinions to create a bigger picture or consensus around knowledge topics (both general and niche), emerging events, audience sentiment, and more. Crowdsourcing refers to such a combination, and has resulted in efforts like the open encyclopedia project Wikipedia generating one of the world's largest knowledge bases, and in previously unimagined solutions to difficult problems posed by companies and other organizations [Tapscott and Williams, 2006].

Even when the "wisdom of the crowds" is not concentrated toward a group of people creating knowledge or solving problems, the information contained in diverse social media updates can be mined and analyzed to derive a more coherent picture of an event or topic than could be found through a single report. In "citizen sensing," information is mined from multiple social media contributions that are related by topic or by geographic location. For example, during events like Occupy Wall Street and the U.S. elections, the Twitris[16] system [Jadhav et al., 2010] provided an interface that could be used to interact with and zoom in to particular sub-events and geographic areas.

Grouping by location can either be carried out through geographic entity recognition methods on the social media texts [Kinsella et al., 2011], or through latitude/longitude proximity measures based on coordinates attached as metadata to social media items (as provided by location-based services such as GPS or triangulation using cell towers and Wi-Fi).

3.1.6 LIMITATIONS WITH SOCIAL SPACES

However, as more and more social websites, communities, and services come online, the lack of interoperation among them becomes obvious: the Social Web mainly consists of sets of single data

---

[15]https://twitter.com/chrismessina/status/223115412.
[16]http://twitris.knoesis.org/.

silos—sites, communities, and services that cannot interoperate with each other, where synergies are expensive to exploit, and where reuse and interlinking of data is difficult and cumbersome.

The main reason for this lack of interoperation is that for most social applications, communities, and domains, there are still no common standards for knowledge and information exchange and interoperation available. (We will refer to these as social spaces in later chapters.)

3.2   CONCLUSION

In this chapter, the first of two parallel developments in the Web's evolution was described: the Social Web. Before we go on to describe how an improved Social Web (with more meaningful and useful semantics) is being formed, we must first define the Semantic Web and some of the possibilities it brings with it.

CHAPTER 4

The Semantic Web

4.1 INTRODUCTION

As mentioned in Chapter 1, the Web was originally conceived to be more than just an information resource: by incorporating machine-interpretable metadata in the Web, computers could leverage this data to help users carry out their online activities. A *Scientific American* article from Berners-Lee, Hendler and Lassila [Berners-Lee et al., 2001] defined the Semantic Web as *"an extension of the current Web in which information is given well-defined meaning, better enabling computers and people to work in cooperation."*

The word "semantic" stands for "the meaning of," and therefore the Semantic Web is one that is able to describe things in a way that computers can better understand.

The last ten years have seen major efforts going into the definition of the foundational standards supporting data interchange and interoperation, and currently a quite well-defined Semantic Web technology stack exists, centered on defining metadata and vocabularies.

However, the number of vocabularies that have actually achieved wide deployment is low. Successful examples include schema.org for marking up website content for search engines, RSS for the syndication of information, FOAF, for expressing personal profile and social networking information, Dublin Core, for expressing information about publications and documents, and SIOC, for interlinking communities and online conversations.

These vocabularies share a joint property: they are small, but at the same time vertical Ð i.e., they are a part of many different domains. Each horizontal domain (e.g., e-health) would typically reuse a number of these vertical vocabularies, and when deployed the vocabularies are able to interact with each other. (We shall note this later when we look at some of the datasets in the "Linked Open Data Cloud.")

Some popular horizontal vocabularies also exist, for example, GoodRelations which is used for product descriptions and e-commerce.

4.1.1 FROM SYNTAX TO SEMANTICS

The structural and syntactic Web put in place in the early nineties is still much the same as what we use today: resources (webpages, files, etc.) connected by semantically untyped hyperlinks. By untyped, we mean that there is no easy way for a computer to figure out what a link between two pages means—what are the semantics of the relationship between the pages. For example, on the UEFA football website (UEFA is the Union of European Football Associations), there are hundreds of links to the various organizations that are registered members of the association, but there is nothing explicitly saying that the link is to an organization that is a "member of" UEFA or what type of organization is represented by the link. On a professor's work page, she may link to many papers that she has written, but the link itself does not say that she is the author of those papers or that she wrote such-and-such when she was visiting at a particular university.

In fact, the Web was envisaged to be much more (Berners-Lee, 1989), as one can see from the image below (Fig. 4.1) which is taken from Tim Berners-Lee's original outline for the Web in 1989, entitled "Information Management: A Proposal." In this, all the resources are connected by links describing the type of relationships, e.g., "wrote," "describe," "refers to," etc. This was a precursor to the Semantic Web.

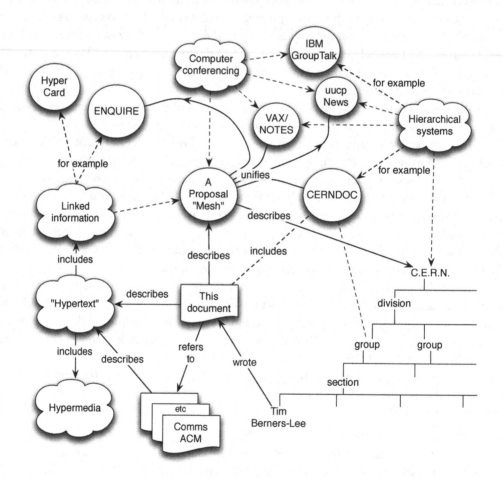

Figure 4.1: The "Mesh" proposal from Tim Berners-Lee.

4.1.2 A GREAT BIG GRAPH OF METADATA AND VOCABULARIES

The Semantic Web consists of metadata that is associated with web resources, and there are associated vocabularies or "ontologies" that describe what this metadata is and how it is all related to each other. We will describe the connection between metadata and vocabularies by example.

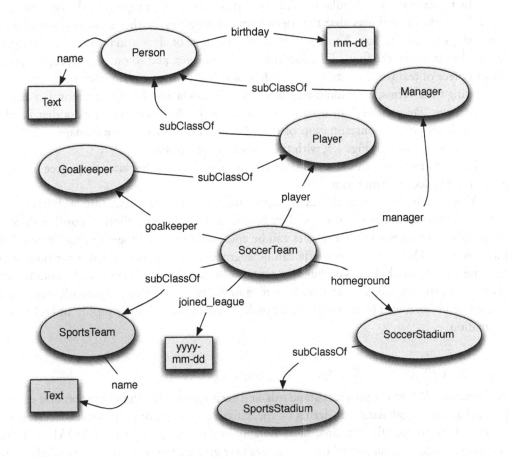

Figure 4.2: An example vocabulary based on soccer.

Let us imagine we wanted to create a vocabulary of terms (and how they relate to each other) around the domain of soccer, so that we can provide metadata from a soccer website about a particular football league. We may want to describe a soccer team (see Fig. 4.2), and the fact that it has various properties, like having a manager, a goalkeeper, players, a homeground, and so on. It may also be that we want to describe more information about the soccer stadium that acts as the team's homeground, for example, where it is, what the crowd capacity is, and so on.

Fortunately, we can also reuse attributes from more generic terms in other vocabularies that are related to our soccer example. A goalkeeper, player, and manager are all types of people, and we can reuse a more generic ontology about people that defines their essential characteristics (name, age, birthdate, etc.). Similarly, teams and stadiums are features of many other types of sports games, and we could link to and reuse terms from a sports ontology.

The terms in these vocabularies take two forms: classes and properties. We typically assign "classes" to entities or things that themselves have "properties," either standalone attributes or relationships to other things. A soccer team would be a type of class, and it would have properties like the date it joined the football league, the team's manager, and so on. These properties could link to a piece of text (e.g., a date) or to another class (e.g., a soccer team manager).

Using these terms, we could then publish typed metadata about soccer teams. For example, we could have "Polar Bears," an instance/instantiation of the SoccerTeam class that is related to "Adam Marcsson," an instantiation of the Manager class through the manager relationship property. This is shown in Fig. 4.3, with the unused terms omitted.

The metadata from this site can then be exchanged with other sites that agree to use this common set of soccer team terms.

As we can imagine from this example, the main power of the Semantic Web lies in interoperability, and combinations of vocabulary terms. Increased connectivity is possible through a commonality of expression; vocabularies can be combined and used together: e.g., a description of a book using Dublin Core metadata can be augmented with specifics about the book author using the FOAF vocabulary. Vocabularies can also be easily extended (through modules, etc.). Through this, true intelligent search with more granularity and relevance is possible: e.g., a search can even be personalized to an individual by making use of their identity profile and relationship information.

4.1.3   ISSUES WITH VOCABULARY CREATION

The Semantic Web effort provides standards and technologies for the definition and exchange of metadata and vocabularies. In terms of vocabularies, there have been some successes in the Semantic Web in specific domains, most notably in the social/community FOAF and SIOC vocabularies which are supported through active user groups; the schema.org effort led by major search engines; in the GeneOnt, HLCS WG, and Open Biomedical Ontologies; and in industry related projects ranging from the oil and gas industry to e-business domains.

However, there are still many different vocabularies on the Semantic Web that could be used together and remain disconnected, since most developers or metadata creators are either unsure about how they can be related to each other or have just not come across them (unless they find them using services such as SchemaWeb or LOV). Also, many people are developing similar or related vocabularies in parallel, unaware of each other's existence.

Since creating a Semantic Web vocabulary for an entire domain can be an arduous process, doing so in isolation can make this a daunting task for many developers, and it is because of this

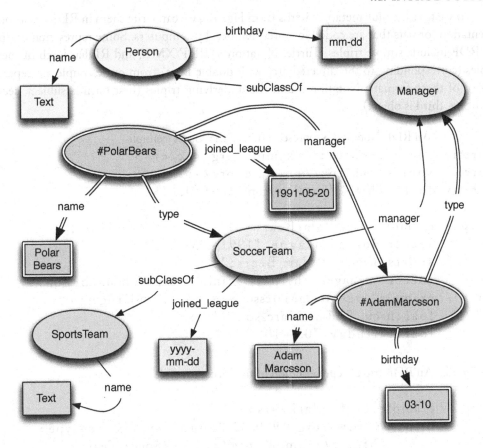

Figure 4.3: Metadata (double lines) and vocabularies (single lines) combined in a soccer example.

that many vocabularies cannot receive critical feedback and thereby do not reach critical mass. Also, it is unclear how to establish a Semantic Web vocabulary in a specific domain, or what the process should be to arrive at a community consensus.

4.1.4   REPRESENTATION FORMATS

We have seen in Fig. 4.3 that the Semantic Web consists of a big graph of interlinked things. One way to represent this graph is using the Resource Description Framework, or RDF. RDF consists of a set of "triples," where things are linked to other things and text strings by a directed arc. We call the originating thing the subject, the arc is called the predicate, and the destination is called the object.

If we take the blue metadata items from Fig. 4.3, we can write them in RDF in various representation formats that are coded for interpretation by computers. Some representation formats for RDF include simple triples, Turtle, Notation 3, RDF/XML, and RDFa. Each of the seven triples (corresponding to the directed arcs with double lines) from our example are represented in two of these formats: Notation 3 and the underlying triples (first term is subject, second is predicate, third is object):

Listing 4.1: An RDF Notation 3 representation of the soccer example

```
@prefix soccer: <http://example.org/soccer#>.
@prefix sports: <http://example.org/sports#>.
@prefix foaf: <http://xmlns.com/foaf/0.1/>.

<http://example.org#PolarBears> a soccer:SoccerTeam;
        soccer:joined_league "1991-05-20;''
        sports:name "Polar Bears;''
        soccer:manager <http://example.org#AdamMarcsson>.
<http://example.org#AdamMarcsson> a soccer:Manager;
        foaf:name "Adam Marcsson;''
        foaf:birthday "03-10.''
```

Listing 4.2: An RDF triples representation of the soccer example

```
<http://example.org#PolarBears>
        <http://www.w3.org/1999/02/22-rdf-syntax-ns#type>
                <http://example.org/soccer#SoccerTeam>.

<http://example.org#PolarBears>
        <http://example.org/soccer#joined_league> "1991-05-20.''

<http://example.org#PolarBears>
        <http://example.org/sports#name> "Polar Bears.''

<http://example.org#PolarBears>
        <http://example.org/soccer#manager>
                <http://example.org#AdamMarcsson>.

<http://example.org#AdamMarcsson>
        <http://www.w3.org/1999/02/22-rdf-syntax-ns#type>
                <http://example.org/soccer#Manager>.
```

```
< http :// example . org#AdamMarcsson >
    < http :// xmlns . com/ foaf /0.1/ name >  "Adam  Marcsson . ''

< http :// example . org#AdamMarcsson >
    < http :// xmlns . com/ foaf /0.1/ birthday >  "03 − 10 . ''
```

4.1.5   SEMANTIC WEB AND SEO

SEO experts have known that adding metadata to their websites can often improve the percentage of relevant document hits in search engine result lists, but it has been difficult in the past to persuade web authors to add metadata to their pages in a consistent, reliable manner (either due to perceived high entry costs or because it is too time consuming). For example, few web authors made use of the simple Dublin Core metadata system, even though the use of DC meta tags can increase their pages' prominence in search results.

However efforts like the Facebook Open Graph Protocol, and the schema.org joint initiative from Google, Bing, Yahoo, and Yandex, have raised awareness of the usefulness of marking up web content with metadata elements.

4.1.6   COMPARISONS WITH MICROFORMATS AND MICRODATA

In parallel with the Semantic Web effort, microformats have been quite successful in bringing semantic metadata to the current Web through a vibrant community centered around a MediaWiki installation, IRC channel, and associated mailing list. Through the microformats community, several open data formats have been created and are currently in use. We will look at some of these in Section 5.8.2.

Microdata was an effort to bring a more standardized effort to embed metadata with HTML5, but in a syntax that would be simpler than RDF and microformats. It is largely being driven by the schema.org initiative, but support for processing microdata has been gradually dropped from some of the main web browsers.

4.2   CONCLUSION

In this chapter, the second of two parallel developments in the evolution of the Web was discussed: the Semantic Web. We next discuss the formation of a Social Semantic Web and some resulting possibilities.

CHAPTER 5

The Social Semantic Web

5.1 INTRODUCTION

As we have seen in Chapter 3, the Social Web has been widely adopted, allowing social interaction and participation through the creation of social spaces. Unfortunately, these social spaces are experiencing limitations in terms of data reuse, interconnectivity, collaboration functionality, and usability. Many act as data silos which restrict various opportunities for added value if they were easily connectable and could be connected to.

Fortunately, key Semantic Web (Chapter 4) technologies and standards are maturing in parallel with the Social Web. The social spaces of the Social Web can be combined with Semantic Web technologies to catalyze the next stage of the Web and to enable new applications in terms of knowledge discovery and data mining. We will describe this convergence in this chapter.

5.2 THE SOCIAL SEMANTIC WEB

This intersection of the Semantic Web and the Social Web is termed the "Social Semantic Web" [Breslin et al., 2009]. The Social Semantic Web aims to overcome some of the aforementioned limitations through a combination of Social Web frameworks with Semantic Web standards, thereby creating a technology platform enabling semantically enhanced social spaces where individuals and communities participate in creating distributed interoperable information.

It is a two-way street: the Semantic Web can help the Social Web and vice versa. The Semantic Web has suffered from a chicken-and-egg problem in the past, whereby it has been difficult to gather semantically rich data for Semantic Web applications to use; however users of the Social Web are creating semantically rich data every second. In the reverse direction, various heterogeneous platforms and Social Web clients can benefit from having interoperable semantic representations of social data to provide integrated views on this data and improved data exchange.

One example is Wikidata, which is bringing more semantic data to the Wikipedia. This can enable contributors to make their work more relevant through better structures, and would give users new querying, browsing, and aggregation possibilities.

5.3 SOME POTENTIAL USES OF THE SOCIAL SEMANTIC WEB

The Social Semantic Web can be a platform for both personal and professional collaborative exchange with reusable community contributions. Through the use of Semantic Web data, search-

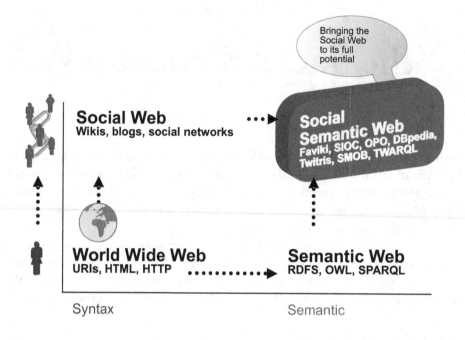

Figure 5.1: The Social Semantic Web.

able and interpretable content is added to existing Social Web collaborative infrastructures, and intelligent use of this content can be made within (and between) these semantically enhanced social spaces—allowing the vision of semantic data on the Web to be realized to its greatest possible advantage.

As mentioned, some typical application areas for the Social Web are wikis, blogs/microblogs, and social networks, but can include any spaces where content is being created, annotated, and shared amongst a community of users. Each of these can be enhanced with machine-readable data to not only provide more functionality internally, but also to create an overall interconnected set of social spaces. This offers a number of possibilities in terms of increased automation and information dissemination that are not easily realizable with current Social Web applications.

1. By providing better interconnection of data, relevant information can be obtained from related social spaces (e.g., through social connections, inferred links, and other references).

2. The Social Semantic Web can allow you to gather all your contributions and profiles across various sites ("subscribe to my brain"), or to gather content from your friend/colleague connections.

3. These semantically enhanced social spaces allow the use of the Web as a clipboard to allow exchange between various collaborative applications (for example, by allowing readers to drag structured information from wiki pages into other applications, geographic data about locations on a wiki page could be used to annotate information on an event or a travel review in a blog post one is writing).

4. Such spaces can help users to avoid having to repeatedly express several times over the same information if they belong to different social spaces.

5. Due to the high semantic information available about users, their interests, and relationships to other entities, personalization of content and interface input mechanisms can be performed, and innovative ways for presenting related information can be created.

6. Fine-grained questions can be answered through such semantically enhanced social spaces, such as "show me all content by people both geographically and socially near to me on the topic of soccer."

7. The Social Semantic Web can make use of emergent semantics to extract more information from both the content and any other embedded metadata (this will be explored later).

There have been initial approaches in collaborative application areas to add semantics to these applications with the aim of adding more functionality and enhanced data exchange—semantic wikis (Semantic MediaWiki), semantic blogs (WordPress with ARC2) and microblogs (SMOB), and semantic social networks (DSSN). These approaches require closer linkages and cross-application demonstrators to create further semantic integration both between and across application areas (e.g., not just blog-to-blog connections, but also blog-to-wiki exchanges).

5.4 INTEGRATING EXISTING SOCIAL SPACES ON THE WEB

A key component of semantically enhanced social spaces is the interconnection of collectively-created, semantically related information, often referred to as "mashups" in Web 2.0 terminology. Currently the term mostly refers to the most basic type of conceivable mashup, where data in static formats from a small number of sources is integrated into novel user interfaces, often based on rigid, application-dependent processing.

One example is Wikitude, an augmented reality app that overlays online geotagged data from Wikipedia on a camera display, contextualized using a phone's current GPS coordinates and compass direction. Even such simple applications provide significant added value to users, but many real-world data sources are far too dynamic, diverse, and complex in structure to allow for such basic treatment.

Wikipedia, for example, incorporates incredible amounts of collaboratively created information on almost every field of human knowledge. As illustrated in existing Wikipedia-powered mashups such as Wikitude, huge added value is to be gained from connecting such information, providing novel surprising views, or from combining it with existing web services and data sources.

Significant community efforts already are bound by manually pursuing similar tasks, e.g., by compiling specific pages which summarize information that is distributed across the encyclopedia, such as the list of countries ordered by their total number of World Cup wins. This manual process, however, is slow and error-prone, leading to incompleteness, replication of data, and internal inconsistencies. By replacing this process with automated methods for knowledge integration, the current manually created replications of data can be superseded by real-time views on the live community-created content of Wikipedia: updating the total number of World Cups won on the page of one country would lead to immediate changes in the list of countries ordered by total World Cup wins.

As content can be processed intelligently inside Wikipedia, it also can be made available to external consumers, making Wikipedia an invaluable free knowledge base to many novel applications. One example of the possible impact of such a transformation of Wikipedia from a content repository to a knowledge base was provided by the DBpedia effort.

5.5   EXTENSION TO FURTHER SOCIAL SPACES

This approach can be extended to further social spaces, yielding unexpected possibilities in creating complex services that go beyond any current mashup. Suitable content is available from innumerable sources on the Social Web: personal and business data on single homepages, photographs on Instagram, entries in private and corporate blogs, messages in forum discussions, product ratings and evaluations on community pages. Many of those sources even provide additional structural information, e.g., relating blog posts by topics, but traditional Social Web technologies cannot achieve an overarching integration as of yet.

On the other hand, the manifold applications brought into being by solving this integration problem can only be imagined. For example, uncovering relationships of blog and forum entries with product profiles can be an instrument for marketing and trend detection. Likewise, end users can connect distributed rating information with product specifications or hotel reviews to make purchasing decisions.

Another range of applications employs methods of personalization and contextualization. For example, photographs taken on a mobile device can be related to time and geographic location, which can again be connected to data from Wikipedia or tourist sites, related photographs from other users, or on-topic news from online feeds. Relevant personalization information stems from users' online identities.

Using technologies such as FOAF and SIOC (more later), online identities can be interlinked with relevant content and user contributions, enabling an overall improved usage experience in an environment of interconnected social websites.

5.6   STANDARD, INTEROPERABLE DESCRIPTIONS OF SOCIAL DATA

The interconnection of existing social spaces would change the nature of content creation and dissemination on the Web. However, information can only be integrated if the structure and semantics of the information is clear. Thus, having standard, interoperable descriptions of data from heterogeneous social spaces is critical to being able to achieve such integration.

This has been well-recognized by the Semantic Web effort, which has emphasized allowing people to create their own vocabularies to describe and share data. The Semantic Web initiative has also resulted in significant achievements, such as standardized vocabulary languages, vocabulary mapping and alignment, semantic data stores, reasoning infrastructures, etc.

The social web requires lightweight semantic structures, or vocabularies. Most of the scenarios discussed above do not require complex descriptions of data. Furthermore, the success of the FOAF and SIOC projects and now schema.org in comparison with other Semantic Web vocabularies provides ample evidence for the value of small vocabularies with community input. These are relatively simple with respect to expressivity, focus on small, well-defined domains, and are supported by active user communities. There are a number of tools that support FOAF and SIOC, making it easy for such vocabularies to be used, evaluated, and extended in a decentralized manner.

Enabling the creation of such niche (vertical) vocabularies and augmenting existing social spaces to enable exchange of information described via such vocabularies opens up a realm of possibilities for applications displaying intelligent behavior and providing enhanced value. Even shallow kinds of semantic structures can enable personalization of websites, clever search and retrieval, intelligent social interaction, etc. Due to the lack of integration, current state of the art has not even scratched the surface of what emergent semantics in social spaces can enable.

5.7   THE LONG TAIL OF INFORMATION DOMAINS

Current approaches to search and author content on the Web are concentrated on just a few, very popular information domains (such as consumer products and reviews, news articles, locations and events, photos, music, and videos). However, due to internationalization and the shift toward an information society, we are facing an increasing quantity of varied information domains. Examples range from the histories of Chinese emperors or the usage of "Daisy Bell (Bicycle Built for Two)" in popular culture to more serious information exchange for enterprise applications and e-government.

Hence, it is important to be able to access and author content not just through a few a priori defined means. Instead, different communities should be able to additionally develop their own tagging systems, taxonomies, terminologies, and ontologies according to their specific needs. In addition, an aim of this book is to show in subsequent chapters how we can mine semantics from content to support authoring, search, and structured information representation in many more information domains.

5.8   SOCIAL SEMANTIC WEB VOCABULARIES

5.8.1   FOAF—FRIEND-OF-A-FRIEND

The Friend-of-a-Friend (FOAF) project[1] was started by Dan Brickley and Libby Miller in 2000 and defines a widely used vocabulary for describing people and the relationships between them, as well as the things that they create and do. It enables people to create machine-readable webpages for people, groups, organizations, and other related concepts. The main classes in the FOAF vocabulary (Fig. 5.2, as illustrated by Dan Brickley[2]) include `foaf:Person` (for describing people), `foaf:OnlineAccount` (for detailing the online user accounts that they hold), and `foaf:Document` (for the documents that people create). Some of the most important properties are `foaf:knows` (used to create an acquaintance link), `foaf:mbox_sha1sum` (a hash over the email address, often used as an identifier for a person and defined as an `owl:InverseFunctionalProperty` to allow smushing/combining instances of the same person from different sources), and `foaf:topic_interest` (used to point to resources representing an interest that a person may have). Other properties from the FOAF specification[3] include `name`, `nick`, `phone`, `homepage`, and `mbox`.

foaf:knows is one of the most used FOAF properties: it acts as a simple way to create social networks through the addition of knows relationships for each individual that a person knows. For example, Bob may specify knows relationships for Alice and Caroline, and Damien may specify a knows relationship for Caroline and Eric; therefore Damien and Bob are connected indirectly via Caroline.

Anyone can create their own FOAF file describing themselves and their social network, using tools such as FOAF-a-matic[4] or FOAF Builder.[5] In addition, the information from multiple FOAF files can easily be combined to obtain a higher-level view of the network across various sources. This means that a group of people can articulate their social network without the need for a single centralized database, following distributed principles enabled by the architecture of the Web.

FOAF can be integrated with any other Semantic Web vocabularies, such as SIOC (described below), SKOS—Simple Knowledge Organization System,[6] etc. People can create their own FOAF document and link to it from their homepage. Aggregations of FOAF data from many individual homepages are creating distributed social networks; this can in turn be connected to FOAF data from larger online social networking sites.

---

[1]http://foaf-project.org.
[2]http://xmlns.com/foaf/spec/images/.
[3]http://xmlns.com/foaf/spec/.
[4]http://www.ldodds.com/foaf/foaf-a-matic.html.
[5]https://github.com/mischat/foafbuilder.
[6]http://www.w3.org/2004/02/skos/.

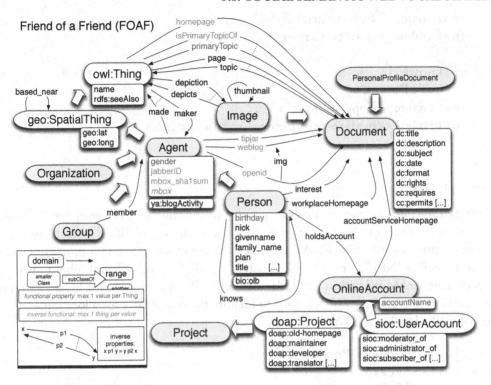

Figure 5.2: Friend-of-a-Friend terms.

The knowledge representation of a person and their friends would be achieved through a FOAF fragment similar to that shown below.

Listing 5.1: A FOAF representation of a person

```
@prefix foaf: <http://xmlns.com/foaf/0.1.> .

<http://www.johnbreslin.com/foaf/foaf.rdf#me> a foaf:Person ;
  foaf:name "John Breslin" ;
  foaf:mbox <mailto:john.breslin@nuigalway.ie> ;
  foaf:homepage <http://www.johnbreslin.org/> ;
  foaf:nick "Cloud" ;
  foaf:depiction <http://upload.wikimedia.org/wikipedia/
  en/b/b6/John_Breslin_in_2008.jpg> ;
  foaf:topic_interest <http://dbpedia.org/resource/SIOC> ;
  foaf:knows [
    a foaf:Person ;
```

```
foaf:name "Sebastian Rios" ;
foaf:mbox <mailto:srios@dii.uchile.cl>
] ;
foaf:knows [
  a foaf:Person ;
  foaf:name "Tope Omitola" ;
  foaf:mbox <mailto:tobo@ecs.soton.ac.uk>
] .
```

5.8.2   HCARD AND XFN

hCard[7] is a microformat used to describe people, organizations, and contact details for both. It was devised by Tantek Çelik and Brian Suda based on the vCard IETF format[8] for describing electronic business cards. Like FOAF, hCard can be used to define various properties relating to people, including "bday" (a person's birth date), "email", "nickname", and "photo", where these properties are embedded within XHTML attributes. The specification for hCard also incorporates the Geo microformat which is used to identify the coordinates for a location or "adr" (address) described within an hCard.

Listing 5.2: hCard for John Breslin

```
<div class="vcard">
<div class="fn">John Breslin</div>
<div class="nickname">Cloud</div>
<div class="org">National University of Ireland Galway</div>
<div class="tel">+35391492622</div>
<a class="url" href="http://johnbreslin.org/">
  http://johnbreslin.org/</a>
</div>
```

XFN (XHTML Friends Network)[9] is another social network-oriented microformat, developed by Tantek Çelik, Eric Meyer, and Matthew Mullenweg in 2003 just before the creation of the microformats community. XFN allows one to define relationships and relationship types between people, for example, "friend", "neighbor", "parent", "met", etc. XFN was also supported through the WordPress blogging platform: when adding a new blogroll link, one could use a form with checkboxes to specify additional metadata regarding the relationship between the blog owner and the person who is being linked to (which is then exposed as metadata embedded in the blog's resulting XHTML).

---

[7]http://microformats.org/wiki/hcard.
[8]http://www.ietf.org/rfc/rfc2426.txt.
[9]http://www.gmpg.org/xfn.

Listing 5.3: An XFN "colleague"-type link to Alexandre Passant
```
<a href="http:// apassant . net /"  rel =" colleague "> Alexandre
    Passant </a>
```

When combined with XFN, hCard provides similar functionality to FOAF in terms of describing people and their social networks. The different types of person-to-person relationships available in XFN allow richer descriptions of social networks to be created as the FOAF vocabulary only has a "knows" relationship. However, FOAF can also be extended with richer relationship types via the XFN in RDF vocabulary[10] developed in 2008 by Richard Cyganiak, or the RELATIONSHIP vocabulary[11] which includes a variety of terms including `siblingOf`, `wouldLikeToKnow`, and `employerOf`.

5.8.3   SIOC–SEMANTICALLY INTERLINKED ONLINE COMMUNITIES

SIOC aims to interlink related online community content from platforms such as blogs, message boards, and other social websites, by providing a lightweight ontology to describe the structure of and activities in online communities, as well as providing a complete food chain for such data. In combination with the FOAF vocabulary for describing people and their friends, and the SKOS model for organizing knowledge, SIOC lets developers link discussion posts and content items to other related discussions and items, people (via their associated user accounts), and topics (using specific "tags", hierarchical categories, or concepts represented with URIs).

As discussions begin to move beyond simple text-based conversations to include audio and video content, SIOC is evolving to describe not only conventional discussion platforms but also new Web-based communication and content-sharing mechanisms. At present, a lot of the content being created on social websites (events, bookmarks, videos, etc.) is being commented on and annotated by others. If you consider such content items to be the starting point for a discussion about the content (similar to a text-based starter post for a thread in a forum or blog), and if the content item being created is done so in a container linked to a user or topic, then SIOC is quite suitable for describing metadata about these content items as well.

Since disconnected social websites could benefit from ontologies for interoperation, and due to the fact that there are a lot of social data with inherent semantics contained in these sites, there is potential for high impact through the successful deployment of a SIOC ontology. The development of SIOC was also an interesting process to explore how an ontology can be developed for and bootstrapped on the Semantic Web. Feedback from the research and development community to the ontology development process was increased through the development of a W3C Member Submission for SIOC.[12]

Some online communities still use mailing lists and message boards as their main communication mechanisms, and the SIOC initiative has also created a number of data producers for

---

[10]http://vocab.sindice.com/xfn.html.
[11]http://vocab.org/relationship/.
[12]http://www.w3.org/Submission/2007/02/.

such systems in order to lift these communities to the Semantic Web. So far, SIOC has been adopted in a framework of over a hundred applications or modules ranging from exporters for major Social Web platforms to applications in neuromedicine research, and has been deployed on about 30,000 websites.[13]

One of the large producers of FOAF, SIOC, SKOS, and Dublin Core (a metadata format for describing both web and physical resources) social semantic data is the Drupal content management system. Drupal has a 6–7% market share amongst content management systems, and the Drupal 7 release has Semantic Web support built-in (generating RDFa data from blog posts, forums, etc.). Some sites that use Drupal 7 and are generating semantic data include energy.gov, london.gov.uk, www.iq.harvard.edu, and software.intel.com. Efforts are currently underway to replace some SIOC terms with types from the schema.org vocabulary (more later) as recommended by four of the major search engines.[14]

An interesting aspect of SIOC is that it goes beyond pure Social Web systems and can be used in other use cases involving the need to model social interaction within communities, either in corporate environments (where there is a parallel lack of integration between social software and other systems in enterprise intranets), or for argumentative discussions and scientific discourse representation (e.g., via the SWAN/SIOC[15] initiative).

The ontology consists of the SIOC Core ontology, an RDF-based schema consisting of 11 classes and 66 properties, and various ontology modules (described in Section 5.8.3).

The SIOC Core ontology defines the main concepts and properties required to describe information from online communities on the Semantic Web. The main terms in the SIOC Core ontology are shown in Fig. 5.3. The basic concepts in SIOC have been chosen to be as generic as possible, thereby allowing us to describe many different kinds of user-generated content.

The SIOC Core ontology was originally created with the terms used to describe web-based discussion areas such as blogs and message boards: namely Site, Forum, and Post (shown in blue in Fig. 5.3). Users create Posts organized in Forums which are hosted on Sites. Posts have reply Posts, and Forums can be parents of other Forums. In parallel with the evolution of new types of social websites, these concepts became subclasses of higher-level concepts shown in yellow—data spaces (`sioc:Space`), containers (`sioc:Container`), and content items (`sioc:Item`)—which were added to SIOC as it evolved. These classes allow us to structure the information in online community sites and distinguish between different kinds of objects. Properties defined in SIOC allow us to describe relations between objects and attributes of these objects. For example:

- The `sioc:has_reply` property links reply posts to the content that they are replying to;

- `sioc:has_creator` (like foaf:maker) links user-generated content to additional information about its author(s); and

---

[13]http://webdatacommons.org/structureddata/2013-11/stats/stats.html.
[14]https://www.drupal.org/node/1784234.
[15]http://www.w3.org/TR/hcls-swansioc/.

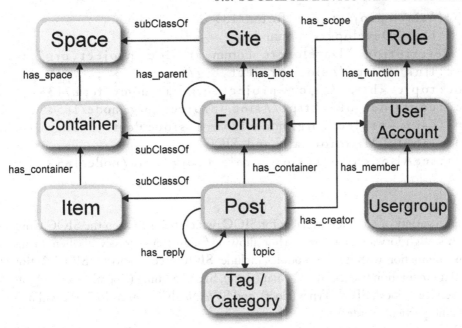

Figure 5.3: The SIOC ontology.

- The `sioc:topic` property points to a resource describing the topic of content items, e.g., their categories and tags.

The high-level concepts `sioc:Space`, `sioc:Container` and `sioc:Item` are at the top of the SIOC class hierarchy, and most of the other SIOC classes are subclasses of these. A data space (`sioc:Space`) is a place where data resides, such as a website, personal desktop, shared file space, etc. It can be the location for a set of Container(s) of content Item(s). Subclasses of Container can be used to further specify typed groupings of Item(s) in online communities. The class `sioc:Item` is a high-level concept for content items and is used for describing user-created content. Usually these high-level concepts are used as abstract classes which other SIOC classes can be derived from. They are needed to ensure that SIOC can evolve and be applied to specific domain areas where definitions of the original SIOC classes such as `sioc:Post` or `sioc:Forum` may be too narrow. For example, an address book, which describes a collection of social and professional contacts, is a type of `sioc:Container`, but it is not the same as a discussion forum.

A sample instance of SIOC metadata from a forum (message board) in Drupal is shown in Listing 5.4. This forum has a title, a taxonomy topic in Drupal, a description, and is the container for one or more posts. More information on the posts can be obtained from the referenced URI (e.g., if it has replies, related posts, who wrote it, etc.).

Listing 5.4: SIOC representation of a message board in RDF/Turtle

```
<http://sioc-project.org/forum/13> a sioc:Forum ;
   dc:title "Developers Forum" ;
   dc:description "Developers Forum at sioc-project.org" ;
   sioc:link <http://sioc-project.org/forum/13> ;
   sioc:topic <http://sioc-project.org/taxonomy/term/13> ;
   sioc:container_of <http://sioc-project.org/node/185> .
<http://sioc-project.org/node/185> a sioc:Post ;
   rdfs:label "Microformats and SIOC" ;
   rdfs:seeAlso <http://sioc-project.org/sioc/node/185> .
```

SIOC modules

There are a variety of modules available for SIOC, which helps to keep the SIOC Core ontology relatively straightforward on its own, while allowing for greater expressivity when the modules are used in conjunction with it. The modules include: SIOC Access (sioce), SIOC Actions (sioca), SIOC Argumentation (siocr), SIOC Chat (siocc), SIOC Mining (siocm), SIOC Quotes (siocq), SIOC Services (siocs), SIOC Types (sioct), and SWAN/SIOC (swansioc). We will now describe these (in no particular order).

A separate SIOC Types module defines more specific subclasses of the SIOC Core concepts which can be used to describe the structure and various types of content of social websites. This module defines subtypes of SIOC objects needed for more precise representation of various elements of online community sites (e.g., `sioct:MessageBoard` is a subclass of sioc:Forum), and introduces new subclasses for describing different kinds of Social Web objects in SIOC. The module also points to existing ontologies suitable for describing full details of these objects (e.g., a `sioct:ReviewArea` may contain Review(s), described in detail using the Review Vocabulary). Examples of SIOC Core ontology classes and the corresponding SIOC Types module subclasses include: sioc:Container (AddressBook, AnnotationSet, AudioChannel, BookmarkFolder, Briefcase, EventCalendar, etc.); sioc:Forum: (ChatChannel, MailingList, MessageBoard, Weblog); and sioc:Post (BlogPost, BoardPost, Comment, InstantMessage, MailMessage, WikiArticle). Some additional terms (Answer, BestAnswer, Question) were also added for question-and-answer-type sites like Yahoo! Answers,[16] whereby content from such sites can also be lifted onto the Semantic Web.

Community sites typically publish web service interfaces for programmatic search and content management services (typically SOAP and/or REST). These services may be generic in nature (with standardized signatures covering input and output message formats) or service specific (where service signatures are unique to specific functions performed, as can be seen in current Web 2.0 API usage patterns). The SIOC Services ontology module allows one to indicate that a web service is associated with (located on) a sioc:Site or a part of it. This module provides a simple way to tell others about a web service, and should not be confused with web service definitions

---

[16]http://answers.yahoo.com.

that define the details of a web service. A `siocs:service_definition` property is used to relate a `siocs:Service` to its full web service definition (e.g., in WSDL, the Web Services Description Language).

The SIOC Actions [Champin and Passant, 2010] module was designed to represent how users of an online community are manipulating the various digital artifacts that constitute the application supporting that community. Use cases include: examining interaction traces to foster group reflexivity and group awareness in collaborative systems; providing a base for digital object memories; and supporting flexible automated reasoning. The main terms in SIOC Actions are `sioca:Action`, `sioca:DigitalArtifact`, `sioca:byproduct`, `sioca:creates`, `sioca:deletes`, `sioca:modifies`, `sioca:object`, `sioca:product`, `sioca:source`, and `sioca:uses`. There is also a set of scripts available[17] to convert Activity Streams,[18] Wikipedia interactions, and Subversion actions into SIOC Actions.

A SIOC Access module was created to define basic information about users' permissions, access rights and the status of content items in online communities. User access rights are modeled using Roles assigned to a user and Permissions on content items associated with these Roles. This module includes terms like `sioce:Status` that can be assigned to content items to indicate their publication status (e.g., public, draft, etc.), and `sioce:Permission` which describes a type of action that can be performed on an object (e.g., a `sioc:Forum`, `sioc:Site`) that is within the scope of a `sioc:Role`.

An argumentation module extension to SIOC has been provided to allow one to formulate agreement and disagreement between SIOC content items. The properties and classes defined in this SIOC Argument module are related to other argumentation models such as SALT[19] and IBIS.[20]

Some related work has also been performed by aligning SIOC with the SWAN ontology for scientific discourse in neuromedicine in the SWAN/SIOC joint initiative,[21] providing a common framework to model online conversations in these communities, from the item level to the conversational layer.

A SIOC Quotes module was also developed to represent quotes in e-mail conversations and other social media content, including the connections between citation blocks and responses.[22]

SIOC Chat provides semantic enrichment of data related to online presence—user accounts, instant messages, chat sessions, and chat topics—which makes it ideal for semantically annotating XMPP messages received from an instant messaging client.

We will also see in the next chapter how SIOC can be used for data mining purposes (using terms defined in the SIOC Mining module).

---

[17]https://svn.liris.cnrs.fr/pchampin/sioca-translator/lib/siocat/.
[18]http://activitystrea.ms.
[19]http://salt.semanticauthoring.org.
[20]http://en.wikipedia.org/wiki/Issue-Based_Information_System.
[21]http://www.w3.org/TR/hcls-swansioc/.
[22]http://slideshare.net/terraces/a-semantic-framework-for-modeling-quotes-in-email-conversations.

5.8.4   OTHER ONTOLOGIES

Some more formats for the semantic modeling of Social Web content have also been created. For example, WIF (Wiki Interchange Format) and WAF (Wiki Archive Format) have been developed by Völkel and Oren [2006] for exchanging and archiving data between different wikis. Also for wikis, the WikiOnt vocabulary was proposed in Harth et al. [2005], and modifications to the SIOC ontology to account for common structures in wikis were described in Orlandi and Passant [2009].

Other application-specific models include SAM [Franz and Staab, 2005] and NABU [Osterfeld et al., 2005] for instant messages, and mle [Rehatschek and Hausenblas, 2007] and SWAML [Fernández et al., 2007] for mailing lists. Both mle and SWAML re-use many terms from SIOC, and the SWAML ontology has been completely integrated with SIOC.

The Online Presence Ontology (OPO)[23] project was initiated for modeling online presence information. Whereas FOAF is mainly focused on static user profiles and SIOC has been somewhat oriented toward threaded discussions, OPO can be used to model dynamic aspects of a user's presence in the online world (e.g., custom messages, IM statuses, etc.). By expressing data using OPO, online presence data can be exchanged between services (chat platforms, social networks, and microblogging services). The ontology can also be used for exchanging IM statuses between IM platforms that use different status scales, since it enables very precise descriptions of IM statuses. OPO and SIOC are also aligned such that semantic descriptions of online presence and community-created content from heterogeneous platforms can be effectively leveraged on the Social Semantic Web.

ActivityStreams[24] is another open format related to the Social Web, whereby feeds of information describing a person's activities can be syndicated (typically using JSON). The initiative was created to address limitations with sharing activity information from the Social Web in Atom and RSS, which was inadequate for expressing the full richness of the original activities.

APML or Attention Profiling Markup Language, was an XML-based format that allowed people to share their own personal "attention profile", similar to how OPML (Outline Processor Markup Language) allows the exchange of reading lists between sites and news readers. APML compressed all forms of attention-related data into a portable file format to provide a complete description of a person's rated interests (and dislikes). The specification is no longer maintained or available.

The Open Graph Protocol[25] was created by Facebook to allow website developers to push content from their websites into the Facebook social graph. It allows metadata about typed content items to be ingested into Facebook, for example, to show information about a song, book, or video that a person has liked or consumed. Initially a much broader set of types was defined

---

[23]http://www.milanstankovic.org/opo/.
[24]http://activitystrea.ms.
[25]http://ogp.me.

(Drink, Game, etc.) and aligned to Semantic Web vocabularies,[26] however these were later deprecated.

Twitter Cards[27] is an effort from Twitter that allows users to "attach rich photos, videos and media experience to Tweets that drive traffic to your website." Similar to (and mapped to concepts from) the Open Graph Protocol, card tags are embedded in a webpage for content such as images, galleries, app downloads, and other media. These cards allow the definition of not just the site owner (using twitter:site to indicate their Twitter handle) but also the creator of the content item itself (using twitter:creator).

The schema.org initiative was launched in 2011 by a group of major search engines to help website developers to markup their data using a common set of schema types. In terms of the Social Semantic Web, some of the main classes are BlogPosting, Comment, Discussion, Person, and some of the main properties include author, comment, commentCount and interactionCount. Mappings to existing RDF vocabularies are also provided.[28]

5.9   CONCLUSION

In this chapter, we have described the concept of the Social Semantic Web—a convergence of two of the parallel evolutions of the Web (Semantic and Social) as outlined in the previous chapters. We describe some of the main advantages made possible by this convergence, and describe some of the main vocabularies that can be used to represent social data from the Social Semantic Web, both RDF-based and other formats.

---

[26]https://github.com/facebook/open-graph-protocol/commit/9cb8522b6b87b08ef93dbc43b3f1e02e57ef54ad.
[27]https://dev.twitter.com/cards/overview.
[28]http://schema.rdfs.org.

CHAPTER 6

Social Semantic Web Mining

6.1 INTRODUCTION

As mentioned in the previous chapter, the Social Semantic Web can be used to bring together data from heterogeneous social websites through common representations and interlinkages. In this chapter, we will describe some extensions that are required when performing web mining on social semantic data, from the perspective of provenance of the data (when data is combined and aggregated from multiple sources), and from the perspective of what the social data actually means in terms of the human interactions, processes, and goals of a community.

We will begin with a focus on provenance, and how it can be used to identify the source/origin of a particular piece of data in a Social Semantic Web, and continue by looking at SIOCM, an extension to existing social semantic vocabularies that is useful in mining and defining the actual usages and policies that are implemented in online communities.

6.2 PROVENANCE

In the previous chapters, we described how web mining of content and user sessions can be used to find clusters of semantically related information, and furthermore, how common structures in social websites can be linked together into a Social Semantic Web. In this section, we look at one of the resulting problems: as sites become more interlinked and information flows from one network to another, how do we track where this information originally came from, and how reputable is it?

6.2.1 RUMORS AND DISSIMULATION IN ONLINE SOCIAL NETWORKS

As the Web allows for information sharing, discovery, aggregation, filtering, and flows in unprecedented ways, it also becomes very difficult to identify reliably the original source that produced an item on the Web. This is highly magnified in online social networks where massive volumes of information are exchanged everyday.

In online social networks, information is published from many different sources, and this information is often re-published and modified. This makes it difficult for a recipient to know where a piece of information originated from, whether or not it should be trusted, or what implicit or explicit biases might be attributed to that piece of information. For example, when a user receives a statement or information through social media, the user needs to make an assessment whether the information is an opinion, a fact, or a rumor.

A resulting requirement is how users can ascribe trust to information that is swirling over social media networks. The trustworthiness of such information from the time of creation, through its aggregation, processing, modification, etc., is essential in order for users to ensure the veracity of a particular data item.

Since typical social networks are global in scale, with billions of users, we need objective and scalable mechanisms that can help users to determine the trustworthiness of data. A logical question to ask then will be what mechanisms would we need in order to help an individual user judge whether or not a statement appearing in social media is fact or fiction. Such mechanisms should provide the user with data about the statement regarding its creation together with its evolution, that can help determine what level of confidence to put in that statement. We may use provenance [Moreau and Missier, 2013] to assist with this, which we will now define.

6.2.2   WHAT IS PROVENANCE?

There have been a variety of definitions for provenance. However, a good starting definition for provenance comes from the W3C Provenance Working Group: *"Provenance is defined as a record that describes the people, institutions, entities, and activities involved in producing, influencing, or delivering a piece of data or a thing."*[1] This is a very pragmatic definition of provenance, especially when applied to the Web context. On the Web, provenance can pertain to data, documents, or in general any resource found on the Web, but it can also appear in a resource on the Web that describes the provenance of a real-world object.

The W3C Provenance Working Group was established to tackle the difficult but worthwhile issue of creating standards to define and manage provenance-related data, and says that "provenance is too broad a term for it to be possible to have one, universal definition—like other related terms such as 'process,' 'accountability,' 'causality' or 'identity,' we can argue about their meanings forever (and philosophers have indeed debated concepts such as identity or causality for thousands of years without converging).[2]

In general, provenance consists of a metadata-based record that describes the entities and processes involved in producing/creating and delivering or otherwise influencing/deriving a (physical or digital) resource [Moreau and Missier, 2013]. This would include information about when an item was created (including the initial sources of information used) together with the various forms of evolution it has undergone (for example, any entity or process involved in producing or altering the resulting piece of information). When described in terms of a process, Moreau [Moreau, 2010] states: "the provenance of a piece of data is the process that led to that piece of data."

In order to better understand how to record provenance, it can be useful to relate it to a scenario in the real world. An example of a provenance record for cake baking is shown below.

---

[1]http://www.w3.org/TR/2013/REC-prov-dm-20130430/ (accessed August 2014).
[2]http://www.w3.org/2005/Incubator/prov/XGR-prov-20101214/ (accessed August 2014).
[3]Adapted from http://tw.rpi.edu/web/project/SPCDIS/Key_Concepts/Provenance (accessed August 2014).

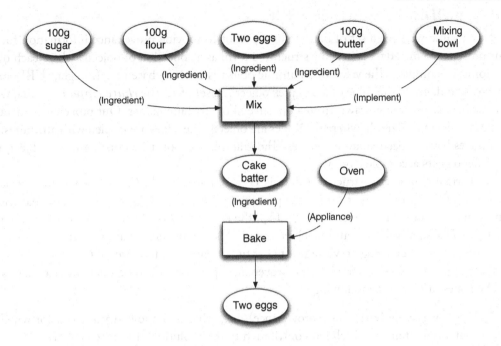

Figure 6.1: A provenance record for cake baking[3].

Since the cake is the item that was created in the end, we can see that there are a sequence of processes followed in its creation (shown here as "Mix" and "Bake"), where the processes also rely on other entities such as ingredients and implements (tools) or appliances.

On the Web, a provenance record can be created by, exchanged between, and processed by computers. Provenance provides a critical foundation for assessing authenticity, enabling trust, and allowing reproducibility. Provenance assertions are a form of contextual metadata and can themselves become important records with their own provenance [Moreau and Missier, 2013].

Provenance therefore provides documentation that is vital to preserving data, determining the quality of the data and its authorship, and is also useful for reproducing and validating results. "It offers the means to verify data products, to infer their quality, to analyze the processes that led to them, and to decide whether they can be trusted" [Moreau, 2010]. For example, by using provenance information it is possible to: enable reproducibility of scientific results [Davidson and Freire, 2008, Gil et al., 2007], or track the authors of particular statements in curated databases [Orlandi and Passant, 2011], or enable reasoning algorithms to make trust assertions about information shared on the Social Web [Artz and Gil, 2007, Carroll et al., 2005].

6.2.3   MODELING PROVENANCE

Let us assume we have a set of data items we want to provide provenance information for and our provenance metadata is in semi-structured form as a collection of objects, where each object is atomic or complex. The value of an atomic object is of some base type (integer, (URI) string, image, sound, etc.). The value of a complex object is a set of $< attribute, object >$ pairs, where an attribute is any string drawn from a universe $\mathcal{A}$ of attribute names. Our provenance metadata can be modeled as a graph, where the nodes are objects, the edges are labeled with attributes, and leaf nodes have associated atomic values. The graph has a root, i.e., a distinguished object, with all other objects accessible from it.

Formally, our semi-structured data is $G = < who \cup < V, E, r, v >>$ where who is the identifier of the actor that performed the provenance operation, V is the set of nodes partitioned into complex and atomic nodes $V = V_c \cup V_a$, the edges are $E \subseteq V_c \times \mathcal{A} \times V, r \in V$ is the root, $v : V_a \to \mathcal{D}$ assigns values to atomic objects, and $\mathcal{D}$ is the universe of atomic values.

So, given an evolving universe $\mathfrak{U}_t$ of G at time t, and a finite subset of G, $\{G_i\}_{1 \leq i \leq M} \subset \mathfrak{U}_T$ from the universe of G at a time T, the provenance questions asked of a data item x, that is one of the nodes of G, are the following:

1. When was x derived (when-provenance), i.e., what is the lowest value of t for which $\mathfrak{U}_t$ contains an item y_a which has contributed to the evolution of x in $\{G_i\} \subset \mathfrak{U}_T$,

2. How was x derived (how-provenance), i.e., what were the first $y, (y = y_1 \ldots y_n)$, in the chain that culminated in the current value of x in $\{G_i\}$,

3. What data was used to derive x (what-provenance), i.e., which $y, (y = y_1 \ldots y_n)$, in $\mathfrak{U}_t$ for $t \leq T$ contributed to the evolution of x in $\{G_i\} \subset \mathfrak{U}_T$,

4. Who carried out the transformation(s) from whence x came (who-provenance), which who was attached to $v : V_x \to \mathcal{D}$.

6.2.4   PROVENANCE OF SOCIAL DATA

Information provenance of social data can be defined as the lineage, management, and the ownership of a piece of information published by an online social platform. Maintaining information provenance in online social networks is a challenging task owing to its huge size and its dynamic nature. When it comes to online social networks, other questions we need to ask when it comes to information provenance management include:

1. What information of a data element to capture and how to model that information?

   Knowing the amount and the levels of provenance information to capture is difficult to judge. There are two main trends of capturing and attaching provenance information: coarse- or fine-grained. Many workflow systems capture provenance information at the coarse-grained document (or file) level, while other transformation systems go down to much finer-grain attribute levels. For many systems, it is better to capture provenance at every level of granularity, as often times many of these systems require a mixture of provenance-capturing techniques.

There are two major approaches to modeling provenance information, and these alternate representations have implications on their cost of recording and the richness of their usages. These two approaches are:

- The inversion method: This uses the relationships between the input data, working backward (hence the name "inversion"), to derive the output data, giving the records of this trace. Examples include queries and user-defined functions in databases that can be inverted automatically or by explicit functions [Widom, 2005]. Here, information about the queries and the output data may be sufficient to identify the source data.

- The annotation method: Metadata of the derivation history of a piece of data are collected as annotations, as well as descriptions about the source data and processes. Here, provenance is pre-computed and readily usable as metadata.

While the inversion method is more compact than the annotation approach, the information it provides is sparse and limited to the derivation history of the data. The annotation method, however, provides more information that includes more than the derivation history of the data and may include the parameters passed to the derivation processes, the post-conditions, etc.

2. How can we store and efficiently access provenance information?

Provenance information can sometimes be larger than the data it describes if the data items under provenance control are fine-grained and the information provided very rich. The inversion method may prove to be more scalable than the annotation method. However, one can reduce storage needs by recording data collections that are important for the operational aspects of the dataset publisher's business.

Provenance can be tightly coupled to the data it describes and located in the same data storage system or even be embedded within the data file. Such approaches can ease maintaining the integrity of provenance, but make it harder to publish and search the provenance data itself. It can also lead to a large amount of provenance information needing to be stored. Provenance can also be stored by itself or with other metadata.

6.2.5   PROVENANCE ON THE WEB

The provenance of information is crucial for information trustworthiness, integration of diverse information assets, etc., on the Web. In 1997, Tim Berners-Lee called for pervasive provenance on the Web:

"At the toolbar (menu, whatever) associated with a document there is a button marked 'Oh, yeah?'. You press it when you lose that feeling of trust. It says to the Web, 'so how do I know I can trust this information?'. The software then goes directly or indirectly back to metainformation about the document, which suggests a number of reasons...."[4]

This pervasive need for provenance can be seen in the Semantic Web stack (Figure 6.2).

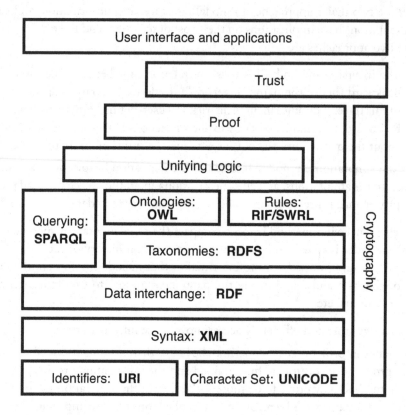

Figure 6.2: The Semantic Web Stack.[5]

The Cryptography layer with the Proof layer help toward laying the foundations of provenance for the Web. The W3C Provenance Incubator Group was set up to provide framework definitions of provenance on the Web. The group carried out the following activities:

- Collected use cases from the Web community that covered varieties of areas emphasizing different provenance needs and uses,

- It organized these use cases according to the different aspects of provenance enumerated above in Section 6.2.2,

---

[4]http://www.w3.org/DesignIssues/UI.html.

[5]Adapted from http://en.wikipedia.org/wiki/File:Semantic_web_stack.svg (accessed September 2014).

- It also developed flagship scenarios that helped coalesce the discussions of the Incubator Group.

The group organized usage requirements and use cases in terms of key provenance categories and dimensions. These categories and dimensions are:

1. **Content**

   These refer to the types of information we need to represent in a provenance record. The dimensions of the content category are:

   - The **object** (or artifact) that we want to make provenance statements about.
   - The **attribution** of that object. This refers to the sources (such as documents, web sites, data, etc.), and/or entities (such as people, organizations, etc.) that contributed to the object in question. The attribution may need to be verified through the use of appropriate authentication mechanisms and may also require some form of anonymization (for privacy reasons).
   - The **process**, i.e., the activities and steps, that was carried out to generate the object.
   - **Versioning** is a useful requirement for a provenance representation. As an object evolves over time, these transformations should be added to the provenance information.
   - Another useful requirement is **decision justification**. The purpose of a justification is to allow decisions to be discussed and understood.

2. **Management**

   This refers to the mechanisms that make provenance information available and accessible. The dimensions of the management category are:

   - **Provenance publication**: Provenance information of an object or objects need to be made available for public consumption. Attendant to provenance publication are the issues of discovery and distribution.
   - **Provenance information accessibility**: Provenance information needs to be easily accessible via some form of query formulation or some other accessibility execution mechanisms.
   - The appropriate levels of **dissemination control** need to be established for the provenance information. This dissemination control mechanisms may include usage policies over the artifact. It may also include licensing information stated by the artifact's creators.
   - Another dimension of the management category is the **scale**, i.e., the size, of the provenance information itself. The size of the provenance records, in some cases, may exceed

the size of the artifacts. There must be trade-offs with respect to the granularity of the provenance information recorded and the level of detail we want to give users of provenance.

3. **Use**

How the provenance information is used is very important. The same provenance information may need to be used differently by different consumers. The dimensions of the usage category include:

- **Understandability**: How to make provenance information understandable to its users is of utmost importance. The providers of the provenance information should make available multiple levels of abstraction as well as multiple perspectives of such provenance information. In order to achieve understandability and usability, it is important to situate the provenance information in the context of the intending users. Here, it is important to combine general provenance information with domain-specific data.

- Provenance **interoperability**: In a Social Web of Data, data and their provenance are obtained from heterogeneous systems. These data may even have different representations. Therefore, interoperability of these heterogeneous provenance data is very important.

- **Comparison** (or equivalence) of artifacts is another important usage dimension. The examination of the provenance records of two seemingly different artifacts may indicate significant commonalities. At the same time, two artifacts that look the same, on the surface, may be quite different when their provenance records are examined.

- One of the uses of provenance information is **accountability**. An object's provenance information can help the object's users to verify if the artifact has gone through the transformations expected of that artifact.

- A good measure of accountability can also be used for **trust**. Provenance information can be used to make trust judgments on a given object.

- Provenance information may be **imperfect**. It may be incomplete or some of it may be incorrect (because of errors).

- Provenance information can be used for **debugging**. Users may want to debug the provenance records in order to locate and probably correct the imperfections that may be noticed in the provenance records.

Integrating Provenance Data into the Architecture of the Web

There are no standard mechanisms to publish and access provenance on the Web. An agreement on a language to use to represent provenance on the Web needs to be accompanied with an agreement or mechanisms of how to find the provenance of a given resource. One way we can do this is by leveraging the HTTP communication mechanisms already existing on the Web. HTTP message exchange can be used to pass provenance between agents: by value, by reference, or mixed.

- Provenance information can be passed by **value** by adding the provenance information directly to the HTTP response or embedded in the HTTP response header. One of the disadvantages of this method is that if the size of the provenance information is large, this may cause a lot of overhead. A good advantage of this method is that the provenance information is never stale and it is as current as the artifact itself.

- Passing provenance information by **reference** involves representing the provenance information as a Web resource and passing its URI to the HTTP response of the artifact itself. An advantage of this method is its very small overhead. A disadvantage of this method is the responsibility it puts on the server to mint new URIs for new provenance records, and also the burden on the server to maintain and keep the provenance Web resources of all the artifacts.

- Provenance information can be passed by **a mixture of value and by reference**. A minimal amount of provenance information can be passed value, while additional provenance records can be represented as Web resources and passed by reference.

Web Provenance Models

1. **voidp** provides the concepts used to describe provenance relationships of data in linked datasets. It builds on and extends the *voiD* ontology and is designed to be lightweight. voiD is an ontology for describing linked datasets. voiD is a vocabulary that attempts to bridge dataset publishers and their users, so that users can find the right data for their tasks more easily using the voiD descriptions of their datasets. This enables users to answer questions such as: "given a set of attributes, which available resources match a desired set of criteria and where can these resources be found?" voidp, the provenance extension to voiD, provides a set of classes and properties that dataset publishers can use to describe the provenance information of the data within their linked datasets.

2. The **Changeset Vocabulary** defines a set of terms for describing changes to resource descriptions. It introduces the concept of a ChangeSet that encapsulates the delta, i.e., the changes, between two versions of a resource description.

3. **PREMIS**. This is a data dictionary to support long-term preservation. It focuses on the provenance of archived, digital objects (files, bitstreams, aggregations), and not on the provenance of the descriptive metadata.

4. **Proof Markup Language (PML)** is a language for representing and sharing explanations generated by various intelligent systems such as theorem provers, task processors, web services, rule engines, and machine learning components, and it provides provenance primitives for annotating and referencing what it refers to as IdentifiedThings. IdentifiedThings

refers to an entity used or processed in an intelligent system, and its properties annotate the entity's properties such as name, description, create date/time, authors, and owner.

5. The **Provenance Vocabulary** was developed to describe provenance of Linked Data on the Web. It is defined as an OWL ontology and it is partitioned into a core ontology and supplementary modules.

6. The Provenance Model (OPM) is used to describe histories in terms of processes, artifacts, and agents. The model was an outcome of a series of Provenance Challenge workshops and was selected as the model to which most of the other provenance vocabularies are mapped to.

6.3   SIOCM

The Web, by connecting people, organizations, and knowledge through their objects-of-interest, is enabling the formation of "object-centered networks" [Breslin and Decker, 2007]. These object-centered networks are inherently social networks, and the people within these networks need to understand what information is accessible in order to know where the gaps are, whom they can connect to, etc.

New tools must be developed to elicit knowledge and useful information from these networks to help people navigate and comprehend these structures. In order to build such tools, one needs to understand the nature of these object-centered networks. One way to do this is to build more intuitive ways of their representations.

The essence of community formation and discovery lies at the heart of these object-centered social networks.

6.3.1   ONTOLOGICAL REPRESENTATION OF ONLINE SOCIAL COMMUNITIES

A community is simply a group of entities that shares a common interest or is involved in an activity or event, and is formally defined as follows [Liu, 2006]:

Given a finite set of **entities** $S = \{s_1, s_2, \ldots, s_n\}$ of the same **type**, a **community** is a pair $C = < T, G >$, where T is the **community theme**, and $G \subseteq S$ is the set of all entities in S that shares the theme T. If $s_i \in G$, s_i is said to be a **member** of the community C.

Several ontologies can be used to represent individuals. The most popular is the Friend-of-a-Friend (FOAF) ontology, the widely used vocabulary outlined in Chapter 5 for describing people, the relationships between them, and their activities (i.e., what they create and do). It enables people to create machine-readable webpages for people, groups, organizations, and other related concepts. As mentioned in that chapter, commonly used classes are foaf:Person (for describing people), and foaf:OnlineAccount (for representing the online accounts users hold). Some of the most important properties are foaf:knows (used to build links and a social network), and foaf:mbox_sha1sum (often used as an identifier for a person).

When it comes to communities, the SIOC ontology provides a vocabulary for describing interlinkages between related online community content from platforms such as blogs, message boards, and other social websites. In combination with the FOAF vocabulary for describing people and their friends, and the SKOS model for organizing knowledge, SIOC lets people link discussion posts and content items to other related items and discussions, to people, and topics.

However, these ontologies lack the appropriate attributes for defining the essence of a community as being comprised of the people in that community, their goals and purposes, and the policies of members' interactions in that community.

FOAF provides the concepts for representing people and their activities, but is missing the concepts that can be used to represent a community's purposes and policies. SIOC, on the other hand, provides the vocabulary for representing an online community, but it is missing the vocabulary for representing a community's purposes, its policies, and the participation levels, e.g., the expertise levels of the community's members. These make it difficult to perform efficient sub-community detection/discovery using these ontologies.

Reflecting, SIOCM is an ontology that has been designed to represent a community's purposes, its policies, and its goals. SIOCM makes extensive use of SIOC and re-uses terms from well-known vocabularies, such as MOAT, and ACL.

SIOCM is made up of the following major modules and classes:

1. Purposes:

   This is comprised of:

   - Tags: given to a particular post by a social entity, and can also be automatically generated by LDA and the text mining of user sessions,
   - Content: this refers to the content of interactions (text, images, sounds, etc.) which are generated during these interactions, and
   - Context: this refers to the context where people interact with one another. A simple example is a thread or a forum on a social network or virtual community. But this could be an XMPP window which holds interactions in form of text messages to other people.

2. Policies:

   This is comprised of:

   - Permissions: this represents the permissions accorded a particular role played by a social entity,
   - Personal Moderation: defines an actor's moderation policies over other actors of the same social entity,
   - Interaction Moderation: this refers to policies regarding interactions on the social entity where the actor is involved, and

- Context Moderation: represents the moderation policies regarding different contexts (such as forums) from a social entity.

3. People:

This is comprised of:

- Social Entity: represents a social entity, such as an actor or a group of actors,
- Group: represents a group or set of actors,
- Person: represents a person or an actor,
- Role: represents the role performed by the social entity, and
- Interaction: refers to interactions between two or more people, and it is formed by several pieces of *content* and belongs to a *context*.

In the next chapter, we will give an application of SIOCM in relation to representing a community's goals and the terms required for this case.

6.4 CONCLUSION

As we create a Social Semantic Web, it is important that we think beyond the common data structures or even what may be latent within the content items and user profiles connected to those data structures. Indeed, the Social Semantic Web can be associated with rich information about what the various social websites are being used for—their purpose, the interactions that take place in communities, and the policies deployed therein. We have presented SIOCM, as an extension to vocabularies like SIOC and FOAF, as one means to represent this richer information. We began the chapter with a description of provenance, which also becomes increasingly important as data is combined across Social Web platforms.

CHAPTER 7

Social Semantic Web Mining of Communities

7.1 INTRODUCTION

Communities are complex organisms. Often the original intent of the community creator (in terms of goals, topics, suitable discussions, imagined members) can diverge significantly as clusters of users gain traction and topics change to new ones of interest. In this chapter, we will look at the evolution of community purpose, and examine a means to detect if the defined community topics (from the overall community or site administrator perspective) are the same as or are significantly different from those actually under discussion by its members (and reflected through their interests).

7.2 PURPOSE EVOLUTION

All virtual communities are different because of their social nature. However, there exist several common characteristics which allow us to classify them (Shummer [2004]). We can find interest communities, purpose communities, and practice communities.

Virtual communities of practice (VCoP) are informal, self-organizing networks of people dedicated to sharing knowledge (Wenger et al. [2002]). An important characteristic of a VCoP is that very often some members help others to answer questions and solve problems. In this way, they share and create knowledge (shared information, good practices, generated software tools, and knowledge bases) (Wenger et al. [2002]).

It is also very important to understand that VCoPs have long-term goals that inspire community owners to create them in the first place. However, to answer if the community members' interests are mainly aligned to the main purpose of the community (its *Leitmotiv*), there is no simple answer. In 2009, Ríos and Aguilera in Ríos et al. [2009] proposed that SNA techniques are not sufficient to fully understand if community members' goals are aligned to those of the VCoP.

SNA analysis is designed to answer questions from a structural viewpoint. This could be a simple question such as: who is the expert or who are the experts in the community? It is measured by studying which nodes are the central ones on a graph (irrespective of whether these central nodes are aligned to the community's main purpose or not). Another interesting question from the SNA viewpoint is to detect sub-communities or subgroups in a graph. But once again, community detection algorithms (clique- or modularity-based) do not consider the semantics of ties among community members. In the following sections, we explain in more detail about how we can understand if community members' comments are aligned to the community owner's (or set of owners') goals.

An important remark is that although community purpose is rather stable (unless community owners decide to change the focus of a web community), community members' goals evolve through time. This means that community members usually have more than one interest/goal when using the VCoP's system, but sometimes they are more interested in one of these and later they might be interested in others. Since we measure goals based on members' posts on the web system, we are able to capture the goal's dynamics and its relation to the community purpose.

7.3   GOALS AS A MEASURE OF PURPOSE ACCOMPLISHMENT

The purpose refers to a community's shared focus on an information need, interest, or service. Every user has his or her own motivations (or interests) to use a specific VCoP, and of course every community of practice can satisfy specific users' needs.

Defining the community's purpose is of major importance, since potential participants can immediately find out about the community's goals (Preece [2004]). However, a VCoP's purpose is not always clear; even worse, it is not clear if all community members are aligned to the same common purpose.

Of course, the key aspect of this work is to consider the purpose evolution analysis as an important tool, but how can we measure purpose? As purpose is something close to the ideas, or underlying motivations of every member, it is not simple to answer this question.

Since purpose is, from the dictionary, "what something is used for," we propose to use goals as a measure of purpose accomplishment. Using this idea, we can measure if a VCoP fulfills a purpose, by measuring how well members' contributions accomplish a set of goals previously defined by the owners, managers, or experts of the community.

Goals definition must be carried out based on interviews or surveys to community experts or administrators (Ríos [2008], Ríos et al. [2006c]). Definition of goals consists of a series of phrases that respond to the question "what is the community for?"

Administrators usually answer different goals depending on their personal interests and viewpoint. Thus, people in charge of experiments should synthesize these sub-goals by validating with administrators or community experts if the goals make sense to them.

This chapter only evaluates the goals from a community administrator's viewpoint. However, similar processes will apply for community members' goals analysis.

Firstly, we need to select a classification or clustering algorithm in order to perform text mining to find interesting patterns. The patterns found can be used by administrators and experts in order to decide how to enhance the community (add new forums, erase forums, find trends, etc.) based on goals fulfillment.

We select a concept-based text mining approach since concept adaptation to goals is straightforward. This approach will be explained in next section.

One last consideration is that this technique allows us to study the goals' fulfillment through time. Therefore, we can show interesting information about how the VCoP's purpose evolves, thus providing useful and objective information to community experts. Without this tool, they only have an intuition of how the community has evolved and whether the information contained in the community forums is truly accomplishing the purpose of the community or not.

7.4   MINING GOALS FROM TEXTS: CONCEPT-BASED TEXT MINING

Concept-based text mining is a data mining approach based on fuzzy sets and fuzzy logic theory. We base our work on Loh's work presented in Loh et al. [2003] and Ríos [2007]. Loh's proposal is to use a *fuzzy reasoning* model to decide whether a concept is expressed by a web document or not. In this way, after application of the reasoning model, we can classify all documents by their concepts. To do so, we compute the degree of possibility that a concept is related to a web document. In our case, web documents are posts in a VCoP's forum.

In this application, we use goals as a way to evaluate the purpose of a VCoP. Therefore, how to introduce goals in the mining process is a key issue. This is why we use concept-based web text mining, as this approach allows us to model goals as concepts, and then based on the goals definition, the algorithm can classify web documents by the accomplishment of such goals. In the following sections, we will talk about goals instead of concepts.

7.4.1   FUZZY LOGIC FOR GOALS CLASSIFICATION

Concept-based text mining uses fuzzy logics' linguistic variables (LV) which are not numbers but words or sentences in natural language. These variables are more complex but less precise. Let u be an LV, we can obtain a set of terms $T(u)$ which cover its universe of discourse U, e.g., $T(taste) = \{sweet, salad, acid, bitter, bittersweet\}$.

In order to use LV for goals classification, we assume that a community's posts can be represented as a fuzzy relation $[Goals \times Posts]$ also called $[G \times P]$. This is a matrix where each row is a goal and every column is a post in the VCoP. To obtain such a matrix, we can rewrite this relation in a more convenient manner in Eq. (7.1) [Loh et al., 2003]. In this expression, we call "Terms" the words that can be used to define a goal and we write "WP" to refer any word inside a webpage. In Eq. (7.1), the symbols "$\times$" and "$\otimes$" represent the fuzzy relation and fuzzy composition respectively.

$$[Goals \times Posts] = [Goals \times Terms] \otimes [Terms \times Posts] \qquad (7.1)$$

As defined above, let P be the total amount of posts in a VCoP, W be the total number of different words among all of these posts, and let G be the total number of goals defined for the community in study. Then we can characterize the matrix $[G \times P]$ by its membership function shown in Eq. (7.2), where $\mu_{G \times P} = \mu_{G \times T} \otimes_{T \times P}$ represents the membership function of the fuzzy composition in Eq. (7.1). The membership values are between 0 and 1.

$$\mu_{G \times P}(x, z) = \begin{pmatrix} \mu_{1,1} & \mu_{1,2} & \cdots & \mu_{1,P} \\ \mu_{2,1} & \mu_{2,2} & \cdots & \mu_{2,P} \\ \vdots & \vdots & \vdots & \vdots \\ \mu_{G,1} & \mu_{G,2} & \cdots & \mu_{G,P} \end{pmatrix} \qquad (7.2)$$

The composition of fuzzy relations is performed using Nakanishi's fuzzy compositional rule in Eq. (7.3). In Eq. (7.3), let $Q(U, V)$ and $Z(V, W)$ be two fuzzy relations which share a common set V. Let $\mu_Q(x, r)$ with $x \in U \wedge r \in V$ and $\mu_z(r, y)$ with $r \in V \wedge y \in W$ be membership functions for Q and Z respectively. Then we can write the compositional rule as shown in Eq. (7.3), where $\bigvee$ is the limited sum $= min(1, x + r)$ and $\wedge$ is the algebraic product $= (x * r)$.

$$\mu_{Q \circ Z} = \bigvee \{\mu_Q(x, r) \wedge \mu_Z(r, y)\} \qquad (7.3)$$

There are several alternatives that can be used to perform the fuzzy composition. Nakanishi et al. [1993] performed a study between six different reasoning models. One important issue that must be considered is that even if some terms are not present in a post, the degree of that post to express a specific goal should not suffer alterations. This is a reason to use Nakanishi's compositional rule. However, other rules could be used as well.

7.4.2   IDENTIFICATION AND DEFINITION OF GOALS

In order to apply the above proposal, we need to begin identifying the relevant goals for the study. To do so, we make use of community administrators' knowledge that identifies which are the most interesting goals to describe the VCoP's purpose. Subsequently, every goal is represented as a list of terms (assuming that a goal is an LV). We used synonyms, quasi-synonyms, antonyms, etc., using also the administrators.

We realize that several important terms are produced by slang words. For example, the word *transformer* in Spanish—"transformador"—is also used as "trafo" or "transf" quite commonly. Another example, the word *amplifier* in Spanish—"amplificador"—is found as "ampli." Thus, human definition is useful to enhance goals definition.

Afterward, we need to define the membership values for the fuzzy relations [*Goals* × *Terms*] and [*Terms* × *Posts*]. We used the relative frequency of terms in a community post to represent the membership values of the matrix [*Terms* × *Posts*].

More difficult is to define [*Goals* × *Terms*] values. We performed this task by asking the community experts to assign these values, for every goal which is the degree that a term has to represent that specific goal. To do so, we compare two terms each time and give a value between 0 and 1. For example, a synonym can receive a value near 1; a quasi-synonym may receive a value somewhere between 0.75 and 0.95; an antonym can be set to 0; etc. This is an indirect method with three experts.

Finally, we obtain the fuzzy relation $\mu_{G \times P}(x, z)$ applying Eq. (7.3). In Table 7.1, we present a column of the matrix $\mu_{G \times P}(x, z)$, which represents the goals classification for post 4235.html from a VCoP (our VCoP of choice is described in the next section). From this table, we can say that post 4325.*html* has a strong relation with goal 1 and goal 2, and almost no relation with goals 3, 4, and 5.

Table 7.1: List of goals and membership values to represent post "4235.htm"

Goals	$\mu_{G \times P}$
Goal 1	0.88
Goal 2	0.72
Goal 3	0
Goal 4	0.12
Goal 5	0.01
....	

7.5 REAL APPLICATION OF COMMUNITY PURPOSE MONITORING

We performed our experiments using the Plexilandia[1] online community. The method described here could be applied to any community/forum (e.g., the boards.ie 10-year dataset in SIOC format.[2] Based on interviews with the administrators of this VCoP and a preliminary study of the community activity, we will now describe the community used.

7.5.1 THE PLEXILANDIA COMMUNITY

Plexilandia is a VCoP formed by a group of people who have come together to build music effects, amplifiers, and audio equipment (in a "Do It Yourself" style). In the beginning, it began as a community to share common experiences in the construction of plexies.[3] Today, Plexilandia has more than 2,750 members who have been sharing and discussing their knowledge about building their own plexies and effects (when we performed our analysis there were over 2,000 members). Also, there are other related topics such as making string instruments, professional audio, and buying/selling parts.

Although, they have a webpage with basic information about the community, most of its members' interactions are produced in the discussion forum area. This community has undergone

---

[1]http://www.plexilandia.cl.
[2]http://data.sioc-project.org.
[3]"Plexi" is the nickname given to the Marshall 1959 amplifier heads that have a clear perspex (Plexiglas) fascia on the control panel with a gold backing sheet showing through, as opposed to the metal plates of later models.

a sustained growth in members' contributions over many years. In the beginning, administration tasks were performed by only one member. Today, this task is performed by several administrators, because the amount of information generated weekly has made it impossible to have the administration tasks fall on just one administrator.

The vision of administrators and experts for the community is based mostly on their experience and time participating in the community. They also have some basic and global measures, such as the total number of posts, online members, etc. However, they do not have any information on members' browsing behaviors, the quality of members publications, and how they contribute to the purpose of the community.

7.5.2   CONCEPT-BASED TEXT MINING APPLICATION

First, we selected data from October 2002 to June 2008 (6 years' worth) and performed text preprocessing to eliminate HTML, Javascript, and other programmed codes.

In order to apply the concept-based text mining approach, we needed to define concepts, or in our case, goals. We defined six different goals with the help of community experts. These were used as inputs and it then took fewer than five minutes to finish the classification process.

7.5.3   ANALYSIS OF RESULTS

Results obtained were included in a web report. This report allows administrators to understand how the purpose of Plexilandia is evolving. The report includes a graph for each forum in the community. This graph represents all goals with a different color. Then, every goal is expressed by its membership value, which means how close that forum is with respect to the VCoP goals. This can be interpreted as degree of accomplishment of a goal by the forum. If a goal has a value near 1, it indicates that the posts in that forum contribute to the accomplishment of that goal (conversely, a value near 0 means that the forum does not help to accomplish a particular goal).

For example, in Figure 7.1, we observe trends in the forum about "amplifiers" and the level of accomplishment for every defined goal. It is possible to observe that: (i) this forum strongly supports goal 1, (ii) in the last few months, this forum has experienced a trend to growth in the accomplishment of goal 2, and (iii) the accomplishment of other goals is much less. This analysis allows us to discover an important conclusion from the administrator's perspective: historically, in the "amplifiers" forum the main topic was amplifiers; however, in the last few months there has been a trend to talk about related topics, such as music effects (e.g., guitar effects). Therefore, this analysis can be thought of as an objective tool to show this emerging situation.

Moreover, the report allows us to identify certain anomalies, such as strange peaks. Administrators have studied these peaks and found that they made perfect sense with particular situations that were occurring in the forums during those months.

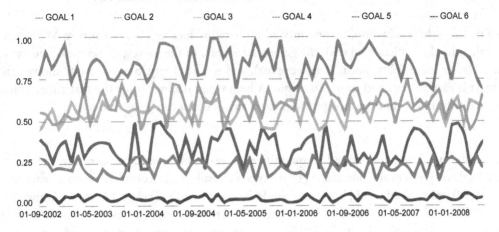

Figure 7.1: Historic goals evolution of the "amplifiers" forum.

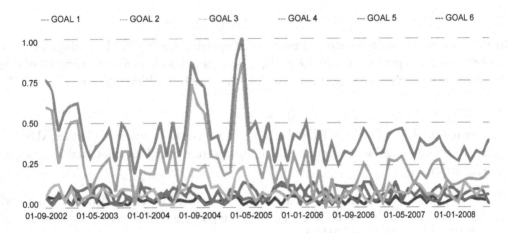

Figure 7.2: Historic goals evolution of the "general" forum.

7.5.4   RESULTS EVALUATION

The main objective of the evaluation was to evaluate the validity of the experimental results. To do so, we performed interviews with community administrators who analyzed and validated the results obtained. The relevance and importance of this evaluation is based on experts' or administrators' knowledge, and their experience about plexies and the community over 6 years. Therefore, they can validate the results, but also they can quantify if a result is expected (just by intuition) or if they have gained an important piece of new knowledge that helps them to better understand the community. In addition, a usability evaluation has been applied to the generated report.

Usability Evaluation

To perform this evaluation, community administrators had to answer a survey. We asked them for their level of satisfaction with the generated report. In the survey, it was measured by: ease of use, ease of learnability, needs help (support), and report clarity. Survey results showed that the report is easy to read and learn, but requires help when reading it for the first time. This is also related with the administrators' previous experiences when using web reports.

Validity of Results

Community administrators had to quantify each result obtained and shown in the report. We used a three-point evaluation scale: (1) unexpected result, i.e., a complete surprise without a clear cause; (2) unexpected result, but cause would be known; and (3) expected result, known cause.

From 14 anomalies detected, 8 were expected results, and 3 represented situations that were completely unexpected. From the 30 identified behavior patterns, 20 represented expected results, and only 2 were completely surprising situations.

7.5.5   APPLYING SIOCM TO STORE GOAL DEFINITIONS

In this section, we shall describe an example of applying the SIOCM ontology presented in Chapter 6 to the representation of the goals, policies, and people in a social network of interest, and the analysis of that network (the network in this case being Plexilandia as described above).

Generating RDF Instances from Plexilandia Database

We adopted the Turtle format of RDF due to its terseness when compared to other RDF formats such as RDF/XML, and we generated RDF instances from the Plexilandia database using a two-pronged effort:

Specify mappings and generating virtual RDF graphs   The Plexilandia database was a MySQL database that had data about users, users' details, their posts, as well as records of their interactions.

R2RML[4] (the RDB-to-RDF Mapping Language) is a language used to map relational data to RDF datasets. It allows relational data to be expressed in the RDF data model. It provides its own grammar which can be used to encode various relational data structures as RDF. We made use of this language to encode the Plexilandia database into RDF.

A snippet of our R2RML structures are given below:

```
# Table fav_cb_puntaje
map:fav_cb_puntaje a d2rq:ClassMap;
d2rq:dataStorage map:database;
d2rq:uriPattern ``fav_cb_puntaje/@@fav_cb_puntaje.id@@;''
d2rq:class vocab:fav_cb_puntaje;
d2rq:classDefinitionLabel ``fav_cb_puntaje;''
```

---

[4]http://www.w3.org/TR/r2rml/.

```
map:fav_cb_puntaje_id a d2rq:PropertyBridge;
d2rq:belongsToClassMap map:fav_cb_puntaje;
d2rq:property vocab:fav_cb_puntaje_id;
d2rq:propertyDefinitionLabel ``fav_cb_puntaje id;''
d2rq:column ``fav_cb_puntaje.id;''
d2rq:datatype xsd:int;
```
.

These R2RML structures above specify how to map one of the tables from Plexilandia, `fav_cb_puntaje`, as a d2rq:ClassMap, and some of its attributes, `fav_cb_puntaje.id`, `fav_cb_puntaje.id,_concepto`, and `fav_cb_puntaje.puntaje`, as d2rq:PropertyBridge.

These mappings were used, together with the data in the Plexilandia database, as input to the D2RQ platform. In the D2RQ platform, we made use of (a) its publishing tool, that publishes the Plexilandia relational data as RDF Linked Data, and (b) the D2R server, a processor that offers a SPARQL endpoint over the mapped relational data.

Using the SIOCM ontology to generate instances   For the network analysis and in order to generate the social network graphs, we developed our own RDB-to-RDF conversion program, written in Java, that used JDBC to connect to the Plexilandia database, getting the MySQL data from the database and converting these to RDF/Turtle. A snippet of the RDF/Turtle schema representation of a Post is shown below:

```
<http://www.plexilandia.cl/foro/post/37610> a sioc:Post ;
def:forum_id "3"^^xsd:string ;
def:post_id "37610"^^xsd:string ;
def:topic_id "4929"^^xsd:string ;
def:topic_poster "793"^^xsd:string .
```

A snippet of the RDF for a social interaction between two members is shown below (LDA represents the Latent Dirichlet Allocation value for this interaction).

```
<http://www.plexilandia.cl/foro/socialinteraction/66/100245>
  a def:SocialInteraction ;
def:lda "0.236321"^^xsd:string ;
def:original_post_user_identifier "100245"^^xsd:string ;
def:reply_post_user_identifier "101262"^^xsd:string .
```

7.5.6   EXTRACTING TOPIC-FILTERED NETWORKS USING SIOCM

Using our semantic representation, we are able to extract a traditional graph representation of Plexilandia. We have comments of this Virtual Community from 2002 to 2010. We count the number of messages that a user sends to another user in the traditional way. We call this the "counting method"; see Fig. 7.3 (left-hand side). In this image, we can observe a graph representation without any filter method. Afterward, using our extended **SIOCM** ontology model, we are able to extract a topic-filtered network—see Fig. 7.3 (right-hand side)—since this information is also stored in our model.

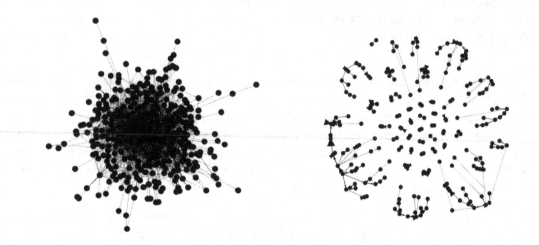

Figure 7.3: Creator-oriented networks using force atlas visualization, with an unfiltered network (traditional) on the left and LDA-filtered network on the right.

The benefit of the above is, for instance, that a filtered network has a density of 0.004 compared to the counting network which has a density of 0.019. Therefore, we obtained a density reduction of about 80% and we keep the quality of traditional methods' results (as shown in Alvarez et al. [2010], L'Huillier et al. [2010]). In addition, since we also store intrinsic topic information, we can also enhance other algorithms' outputs. The SPARQL snippets below show how we got the data for the graph representation:

```
SELECT DISTINCT ?originalPostUser
?replyPostUser ?ldaValue where {?s a
<http://purl.org/net/sioc-sna-dm-uchile/def/SocialInteraction> .
?s <http://purl.org/net/sioc-sna-dm-uchile/def/
   original_post_user_identifier> ?originalPostUser .
?s <http://purl.org/net/sioc-sna-dm-uchile/def/
   reply_post_user_identifier> ?replyPostUser .
?s <http://purl.org/net/sioc-sna-dm-uchile/def/lda> ?ldaValue . };
```

7.6 CONCLUSION

In this chapter, we looked at the purpose evolution of online communities, specifically virtual communities of practice (VCoPs). We think that it is important to point out that community administrators can be unaware of many important aspects if they choose to enhance a social network based solely on either SNA or data mining. It is important to mix both approaches in order to add the semantics layer into community management processes.

We have shown in this chapter that using community experts or administrators combined with a data mining approach can provide much more objective and rich information. This in turn may be used to enhance the virtual community of practice.

In addition, we have proposed a purpose evolution–centric analysis for a VCoP. This is based on goals definition as a key to the data mining analysis for the VCoP. In this way, community experts or administrators can rely on dealing with more objective information.

Finally, we can take experts' knowledge and store it in the SIOCM ontology, which was designed to store fundamental metadata from a VCoP such as purpose and goals. Afterward, the results of concept-based text mining can also be stored in this semantic structure to be used for further querying in more data mining processes.

We have successfully used the purpose evolution analysis in a real VCoP with more than 2,000 members and 6 years of data.

CHAPTER 8

Social Semantic Web Mining of Groups

8.1 INTRODUCTION

As we have seen in the previous chapter, it can sometimes occur that at a community level, the goals of the community may align very well with those of all of its members, or there may be interests in the community that may not align fully with those goals. Another aspect of communities is where groups of users are clustering around certain topics—as fans of innocuous topics or even more serious ones (e.g., terrorism-related). In this chapter, we will look at how we can discover potential groups of topics (e.g., threat-related ones), and the key group members that sustain and propagate these topics. This chapter is based on previous work from Rios et al., with more detail given in Alvarez et al. [2010], L'Huillier et al. [2010], Ríos and Muñoz [2012, 2014].

On the Internet, it is no secret that terrorists and cyber-criminals are looking to discover the means to seek out and connect with peers having similar interests using platforms such as Internet forums, blogs, and social networks, where they can share and comment about their feelings and interests with others who support their cause. In this sense, the Dark Web[1] has brought about the potential to more easily achieve the coordination, propaganda delivery, and other unwanted interactions between extremist groups.

Many researchers have been looking into whether leaders in the Dark Web can be identified by their interactions with other members or not, or if it is possible to detect groups of members who may be potentially dangerous. It is certainly possible to do so purely by using Social Network Analysis (SNA) techniques, however we will show that the results are insufficient to answer these questions unless we add a preprocessing stage which includes higher-order semantic information from texts into these SNA techniques.

The Dark Web consists of Virtual Communities of Interest (VCoIs) [Kosonen, 2009, Porter, 2004]. VCoIs are communities whose members are reunited by their shared interests in a topic, e.g., fan clubs of music artists. Therefore, a key aspect when studying a VCoI is to achieve a complete understanding of what are the main interests of the community. Only then will it be possible to obtain a better insight into a community's social aspects, leading to more accurate identification of the key members (i.e., opinion leaders), sub-groups extraction, density reduction, amongst other benefits.

---

[1]The Dark Web refers to Internet-based forums or platforms that can be used, for example, to discuss radical religious ideas.

Different approaches have been previously developed for social networks, virtual communities [McCallum et al., 2007], and Virtual Communities of Practice (VCOPs) [Bourhis et al., 2005, Probst and Borzillo, 2008, Zhou et al., 2009]. Although Latent Semantic Analysis (LSA) for enhancing Social Network Analysis (SNA) has not been deployed for Dark Web portals before, the application of these techniques may greatly enhance SNA results, and our first work on this matter can be found in Ríos et al. [2009].

This chapter is structured as follows: in Section 8.2, previous work on SNA and text mining for the identification of key members is presented, as well as previous work on the analysis of Dark Web forums. Then, in Section 8.3 the proposed methodology to solve the *topic-based community key members extraction problem* and the main contribution of this methodology is described. In Section 8.4, the experimental setup is detailed, and its results are discussed. Finally, in Section 8.5, the main conclusions of this chapter and future work are presented.

8.2   PREVIOUS WORK

Since the Dark Web is a VCoI, there are different goals associated with their members' objectives [Kim et al., 2010]. In this case, the support of the community through an online forum where its anonymity, ubiquitous nature, and freedom-of-speech exists makes it the perfect environment to share fundamentalism and terrorism propaganda. Therefore, a method to recognize the underlying members' objectives is needed in order to identify threats or security issues. Topic-based SNA [Ríos et al., 2009] is a way to track, to some degree, topics from members' threads, which can be related to members' goals.

In the following, previous work on topic-based SNA will be presented, as well as Dark Web SNA applications.

8.2.1   TOPIC-BASED SOCIAL NETWORK ANALYSIS

Social Network Analysis (SNA) [Wasserman and Faust, 1994] helps us to understand relationships in a given community by analyzing its graph representation. Users are seen as nodes, and relations among users are seen as arcs. Several techniques have been proposed to extract key members [Nolker and Zhou, 2005], classify users according to their relevance within the community [Pfeil and Zaphiris, 2009, Yelupula and Ramaswamy, 2008], discovering and describing resulting sub-communities [Kwak et al., 2009], amongst other applications. However, all these approaches leave aside the meaning of relationships among users. Therefore, analysis based only on the reply of mails or posts to measure a relationship's strength or weakness is not a good indicator.

McCallum et al. (in McCallum et al. [2005, 2007]) described how to determine roles and topics in a text-based social network by building Author-Recipient-Topic (ART) and Role-Author-Recipient-Topic (RART) models. Furthermore, in Pathak et al. [Pathak and Erickson, 2008], a community-based topic-model integrated social network analysis technique (Community-Author-Recipient-Topic model or CART) is proposed to extract communities from an email corpus based on the topics covered by different members of the overall network.

These approaches' novelty is in their use of data mining on text from the social network to perform SNA so as to study roles or to extract subgroups.

8.2.2 SOCIAL NETWORK ANALYSIS ON THE DARK WEB

As described by Xu and Chen (in Xu and Chen [2008], Xu et al. [2006]), the topology of dark networks exhibits different properties when compared with other types of networks, where small-world structures are determined by the information flow properties, characterized by a short average path and a high clustering coefficient. In this sense, social network analysis tools used in other virtual community applications [Ríos et al., 2009] could be useful for analyzing this kind of structure. In their work, the authors combine concepts obtained from communities' administrators into a concept-based text mining process, which allow us to obtain interesting results in terms of purpose accomplishment in a Virtual Community of Practice (VCoP).

In Reid et al. [2005], Web forums were analyzed in order to determine whether a given community has been involved with a terrorist presence by using automated and semi-automated procedures to gather information and analyze it. Other applications, such as the one proposed by Zhou et al. (in Zhou et al. [2005]), aims to capture data from domestic Web forums and use social network mapping to identify their structure and cluster affinities. Finally, in Zhang et al. [2009], a complete framework on how to build a portal of Dark Web forums is presented, where data collection and integration, as well as its visualization and open access, are the main contributions of the authors.

Previous work on key-members identification on the Dark Web [Xu and Chen, 2005] describes how several centrality measures can be used to identify different key members of a social network. Here, the degree, betweenness, and closeness measures have been described as tools to characterize the key members of a given social network.

Latent Semantic Analysis has been applied previously in different Dark Web applications, such as Bradford [2006] where Latent Semantic Indexing was used to link nodes to certain topics in the network construction. Furthermore, in Ng and Yang [2009], a Latent Dirichlet Allocation (LDA)-based web crawling framework was proposed to discover different topics from Dark Web forum sites, but no further social network analysis applications were made with these findings thus, leaving undone the measurement of many social aspects from the Dark Web's social structure.

In terms of text mining in Dark Web forums [Abbasi et al., 2008], applications have been oriented to model assessment and feature selection in order to improve the classification of messages that contain sensitive information on extremists' opinions and sentiments. Also, authorship analysis [Abbasi and Chen, 2005] has been previously applied in order to analyze the groups' authorship tendencies, and more importantly, tackling the anonymity problem associated with these types of virtual communities.

8.3   METHODOLOGY FOR GROUP KEY MEMBER DISCOVERY

The main question of this chapter is how to enhance key-members discovery based on a topic-based social network structure. The first step is to obtain a reduced/filtered representation of the inner social community. However, this representation must be created in such way that information contained therein can be used to discover threats or homeland security issues (it might also be used to understand other non-threatening subjects). The second step is to apply a core members' algorithm, like HITS [Kleinberg, 1999], to obtain the key members of a Dark Web community based on threatening topics. As a result, we obtain a ranking for the complete community of members where for each threatening topic, we are able to reveal members that can be considered as key experts on that topic.

We must mention the difference between a key member and a highly radical member, since key members have several characteristics that define them. Firstly, a key member may be a highly radical member in a given topic or not. Secondly, he/she may increase the interaction in the community because he/she points out interesting messages, which produces replies from different levels of members. We will focus on key members of any kind, and afterward, depending on the topic, it would be possible to classify him/her as a radical member or not. Of course, it is possible to automate the process of radical member discovery, however more work is needed to avoid subjectivity. Furthermore, accusing or adding such label to a person is not a simple matter and its implications are far beyond this work.

We defined a key member as a person totally aligned with the VCoI's goals and topics. Thus, they are producing content which is very relevant toward satisfying other members' interests. The only way to measure a key member as defined is using a hybrid approach of SNA combined with latent semantic-based text mining. In this way, we can discover threatening topics, and perform specific analysis on each one.

8.3.1   BASIC NOTATION

Let us introduce some concepts. In the following, let $\mathcal{V}$ be a vector of words that defines the vocabulary to be used. We will refer to a word w as a basic unit of discrete data, indexed by $\{1, ..., |\mathcal{V}|\}$. A post message is a sequence of S words defined by $\mathbf{w} = (w^1, ..., w^S)$, where w^s represents the sth word in the message. Finally, a corpus is defined by a collection of $\mathcal{P}$ post messages denoted by $\mathcal{C} = (\mathbf{w}_1, ..., \mathbf{w}_{|\mathcal{P}|})$.

A vectorial representation of the posts corpus is given by TF-IDF= $(m_{ij}), i \in \{1, ..., |\mathcal{V}|\}$ and $j \in \{1, ..., |\mathcal{P}|\}$, where m_{ij} is the weight associated to whether a given word is more important than another one in a document. The m_{ij} weights considered here are defined as an improvement on the *tf-idf* term [Saltón et al., 1975] (*Term frequency times inverse document frequency*), defined by

$$m_{ij} = f_{ij}(1 + sw(i)) \times log\left(\frac{|\mathcal{C}|}{n_i}\right) \tag{8.1}$$

where f_{ij} is the frequency of the ith word in the jth document, $sw(i)$ is a factor of relevance associated with word i in a set of words, and n_i is the number of documents containing word i. In this case, $sw(i) = \frac{w_{post}^i}{|\mathcal{P}|}$, where w_{post}^i is the frequency of word i over all documents, and $|\mathcal{P}|$ is the total amount of posts. The *tf-idf* term is a weighted representation of the importance of a given word in a document that belongs to a collection of documents. The *term frequency* (TF) indicates the weight of each word in a document, while the *inverse document frequency* (IDF) states whether the word is frequent or uncommon in the document, setting a lower or higher weight respectively.

8.3.2   TOPIC MODELING

A topic model can be considered as a probabilistic model that relates documents and words through variables which represent the main topics inferred from the text itself. In this context, a document can be considered as a mixture of topics, represented by probability distributions which can generate the words in a document given these topics. The inferring process of the latent variables, or topics, is the key component of this model, whose main objective is to learn from text data the distribution of the underlying topics in a given corpus of text documents.

One main topic model is Latent Dirichlet Allocation (LDA) [Blei et al., 2003]. LDA is a Bayesian model where latent topics of documents are inferred from estimated probability distributions over the training dataset. The key idea of LDA is that every topic is modeled as a probability distribution over the set of words represented by the vocabulary ($w \in \mathcal{V}$), and every document as a probability distribution over a set of topics ($\mathcal{T}$). These distributions are sampled from multinomial Dirichlet distributions.

The advantage this method has over a concept-based approach is that it is an automated process—it only needs experts to provide a name for the topic discovered by the algorithm.

For LDA, given the smoothing parameters β and α, and a joint distribution of a topic mixture θ, the idea is to determine the probability distribution to generate from a set of topics $\mathcal{T}$, a message composed of a set of S words w ($\mathbf{w} = (w^1, ..., w^S)$),

$$p(\theta, z, \mathbf{w}|\alpha, \beta) = p(\theta|\alpha) \prod_{s=1}^{S} p(z_s|\theta)p(w^s|z_s, \beta) \tag{8.2}$$

where $p(z_s|\theta)$ can be represented by the random variable θ_i, such that topic z_s is presented in document i ($z_s^i = 1$). A final expression can be deduced by integrating Eq. (8.2) over the random variable θ and summing over topics $z \in \mathcal{T}$. Given this, the marginal distribution of a message can be defined as follows:

$$p(\mathbf{w}|\alpha, \beta) = \int p(\theta|\alpha) \left(\prod_{s=1}^{S} \sum_{z_s \in \mathcal{T}} p(z_s|\theta) p(w^s|z_s, \beta) \right) d\theta \tag{8.3}$$

The final goal of LDA is to estimate previously described distributions to build a generative model for a given corpus of messages. There are several methods developed for making inferences over these probability distributions, such as variational expectation-maximization [Blei et al., 2003], a variational discrete approximation of Eq. (8.3) empirically used by Xing and Girolami [2007], and a Gibbs-sampling Markov-chain Monte Carlo model efficiently implemented and applied by Phang and Nguyen [2008].

8.3.3 NETWORK CONFIGURATION

To build the social network, the members' interactions must be taken into consideration. In general, members' activities are followed according to their participation on the forum. Likewise, participation appears when a member posts in the community. Because the activity of the VCoI is described according to members' participation, the network is configured according to the following: nodes will be the VCoI members, and arcs will represent interactions between them. How to link the members and how to measure their interactions to complete the network is our main concern.

Here, we will describe two VCoIs' network representations according to the following replying schema of members:

1. **Creator-Oriented Network**: When a member creates a thread, every reply will be related to him/her.

2. **Last Reply–Oriented Network**: Every reply of a thread will be a response to the last post.

In Fig. 8.1, the latter two approaches of network conversion of the forum are presented. In Fig. 8.1, arcs represent members' replies, and nodes represent the users who made the posts. In our first approach, the weight of an arc will be a count of how many times a given member replies to others.

In order to consider the replies of members made according to the community's purpose (for any of these configurations), and to filter noisy posts, a topic-based message reduction is performed. According to previous network configurations, the topic-based filtering method is used in order to remove all replies that are not in accord with the posts' topic. Here, by using LDA, a list of keywords for each topic concept is determined, and later, used to build the final version of the social network.

8.3.4 TOPIC-BASED NETWORK FILTERING

The previous chapter described a method to evaluate a community's goals accomplishment. In this chapter, we will use this approach to classify members' posts according to a VCoI's goals.

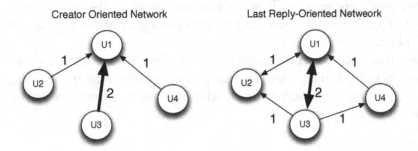

Figure 8.1: Two different network models to represent a given forum interaction.

These goals are defined as a set of terms, which are composed of a set of keywords or statements in natural language.

The idea is to compare with the Euclidean distance between two members' posts. If the distance is over a certain threshold ϕ, an interaction will be considered to exist between them. We support the idea that this will help us to avoid irrelevant interactions. For example, in a VCoI with k goals (or topics), let TB_j be a post by user j that is a reply to a post by user i (TB_i). The distance between them will be calculated with Eq. (8.4):

$$d_m(TB_i, TB_j) = \frac{\sum_k g_{ik} g_{jk}}{\sqrt{\sum_k g_{ik}^2 \sum_k g_{jk}^2}} \qquad (8.4)$$

where g_{ik} is the score of topic k in a post by user i. It is clear that the distance exists only if TB_j is a reply to TB_i. After that, the weight of arc $a_{i,j}$ is calculated according to Eq. (8.5):

$$a_{i,j} = \sum_{\substack{i,j \\ d_m(TB_i,TB_j)}} d(TB_i, TB_j) \qquad (8.5)$$

We used this weight in both configurations previously described (Creator-Oriented and Last Reply–Oriented). Afterward, we applied HITS [Kleinberg, 1999] to find the key members on the different network configurations.

8.3.5 NETWORK CONSTRUCTION

Algorithm 1 presents the pseudo-code of how the graph $\mathcal{G}_c = (\mathcal{N}, \mathcal{A})$ is built by using the Creator-Oriented Network, and algorithm 2 presents how it is used in the Last Reply–Oriented Network construction.

Algorithm 1 shows the initial definition of the `TF-IDF` matrix, as well as the topic extraction method from the posts' corpus by using LDA, with a given set of parameters $\{k, n, \ , \tau\}$. Afterward, the `TF-IDF` matrix is multiplied by the `Topic` matrix in order to clean up the overall corpus's vectorial representation. Then, for each post creator i, the arc weight $a_{i,j}$ is increased according to the number of repliers j, where their messages' distance is greater than or equal to the threshold θ.

Algorithm 2 also shows the initial definition of the TF-IDF matrix, as well as the Topic matrix, and its multiplication. Then, for each post's username i, the arc weight $a_{i,j}$ is increased according to its number of direct repliers j whose messages' distances are greater than or equal to a threshold θ.

Data: $\mathcal{V}, \mathcal{P}, k, n, \alpha, \beta$
Result: $\mathcal{G}_c = (\mathcal{N}, \mathcal{A})$
TF-IDF$[|\mathcal{P}| \times |\mathcal{V}|] \leftarrow$ Build TF-IDF matrix using Eq. (8.1) ;
Topic$[k \times \mathcal{V}] \leftarrow$ Build Topic matrix using LDA(k, n, α, β) (section 8.3.2);
$TB[|\mathcal{P}| \times k] \leftarrow$ TF-IDF $\otimes$ Topic^T;
Initialize $\mathcal{N} = \{\}, \mathcal{A} = \{\}$;
foreach $i \in \mathcal{P}$ **do**
$\quad \mid \quad \mathcal{N} \leftarrow \mathcal{N} \cup i$;

foreach $i \in \mathcal{P}.creator$ **do**
$\quad$ **foreach** $j \in \mathcal{P}, i \neq j$ **do**
$\quad\quad$ **if** $d_m(TB_i, TB_j) \geq \theta$ **then**
$\quad\quad\quad a_{i,j} \leftarrow a_{i,j} + 1$;
$\quad\quad\quad \mathcal{A} \leftarrow \mathcal{A} \cup a_{i,j}$;

return $\mathcal{G}_c = (\mathcal{N}, \mathcal{A})$;

Algorithm 1: Creator-Oriented Topic-Based Network

8.4   EXPERIMENTAL SETUP AND RESULTS

The proposed methodology was applied in the `Ansar1` English language–based forum, available on the Dark Web portal,[2] for which examples of the topic extraction methodology, network construction, and key-members identification using the HITS algorithm were determined.

Topics extracted were evaluated according to the compactness of each topic, by the Average Shortest Distance (ASD) among the top 30 words in the topic [Yang et al., 2009]. The ASD between two keywords w_i and w_j is determined by:

$$\text{ASD}(w_i, w_j) = \frac{max\{log(N_i), log(N_j)\} - log(N_{i,j})}{log(|\mathcal{P}| - min\{log(N_i), log(N_j)\})} \qquad (8.6)$$

where N_i is the number of posts that contains w_i, $N_{i,j}$ is the number of posts that contain both of the keywords, and $|\mathcal{P}|$ is the total number of posts in the corpus.

[2]`http://128.196.40.222:8080/CRI_Indexed_new/login.jsp`.

Data: $\mathcal{V}, \mathcal{P}, k, n$
Result: $\mathcal{G}_r = (\mathcal{N}, \mathcal{A})$
TF-IDF$[|\mathcal{P}| \times |\mathcal{V}|] \leftarrow$ Build TF-IDF matrix using Eq. (8.1) ;
Topic$[k \times |\mathcal{V}|] \leftarrow$ Build Topic matrix using LDA(k, n, α, β) (section 8.3.2);
$TB[|\mathcal{P}| \times k] \leftarrow$ TF-IDF $\otimes$ Topic^T;
Initialize $\mathcal{N} = \{\}, \mathcal{A} = \{\}$;
foreach $i \in \mathcal{P}$ **do**
  $\quad \lfloor \quad \mathcal{N} \leftarrow \mathcal{N} \cup i$;
foreach $i \in \mathcal{P}$ **do**
  $\quad$ **foreach** $j \in \mathcal{P}.reply(i), i \neq j$ **do**
  $\quad\quad$ **if** $d_m(TB_i, TB_j) \geq \theta$ **then**
  $\quad\quad\quad$ $a_{i,j} \leftarrow a_{i,j} + 1$;
  $\quad\quad\quad$ $\mathcal{A} \leftarrow \mathcal{A} \cup a_{i,j}$;

return $\mathcal{G}_r = (\mathcal{N}, \mathcal{A})$;
Algorithm 2: Reply-Oriented Topic-Based Network

As a benchmark, the direct graph construction and HITS evaluation was performed over the whole corpus of posts without topic filtering. Then, the representation of the overall results was compared against a month-based analysis of the communities' most replied-to members on both network representations. For this, the top five keymembers were tracked in the overall evaluation of the networks.

8.4.1 RESULTS AND DISCUSSION

In this section, relevant results are presented and discussed accordingly. Firstly, some topic-extraction results are presented as well as a brief analysis of the main topics extracted. Secondly, different network representations are given as well as a benchmark network evaluation. Finally, some HITS algorithm results and discussion are presented for a given topic-based network.

Topics Extraction
We used 14 months of data (from December 2008 to January 2010) from the aforementioned portal. Posts were created by 376 members[3] and topics were extracted from 29,057 posts $\mathcal{P}$ and 103,791 words in the vocabulary $\mathcal{V}$ by using a C++ Gibbs sampling-based implementation of LDA,[4] previously described in section 8.3.2. By using Eq. (8.6) [Yang et al., 2009], a representative number of topics was determined, by evaluating the ASD number for $k \in [10, 50]$, where 47 was the number of topics with the lowest ASD.

---

[3]Members were anonymized for the purposes of this work.
[4]http://gibbslda.sourceforge.net/.

The overall list of resulting topics was revisited by visual inspection, and we proposed descriptive names for different topics. Furthermore, different topics were grouped into concept categories, for which social network analysis was taken into further consideration. In this context, a total of 11 topics extracted were not assigned to a descriptive name or concept because of its lack of interpretability (and the sparseness of its words' concepts).

In Table 8.1, the overall concept proposed is "Local Conflicts" as its main topics are related to terrorist and counter-terrorism activities over different localities, such as Russia, Iraq, Somalia, India, and Afghanistan. The overall concept proposed for the results shown in Table 8.2 is "War on Terror," as its main topics are related to U.S. activities and activities toward terrorism activities, where the proposed topic names are "War," "American Soldier," "Imprisonment," "Obama," and "Military Operations." Finally, as shown in Table 8.3, the overall concept proposed is "Recruiting" where topics such as "Religious Propaganda," "Religious Conflicts," "*Ansar*[5] Propaganda," "Family," and "Ideology" are included.

Table 8.1: Ten most relevant words with their respective conditional probabilities for five topics associated with the "Local Conflicts" concept from the `Ansar1` forum

Topic 1 **"Eastern Europe"**	Topic 8 **"Iraq Conflicts"**	Topic 13 **"Somalia Conflicts"**	Topic 17 **"India Conflicts"**	Topic 19 **"Afghanistan Conflicts"**
russian (0.0289)	iraq (0.0885)	somalia (0.0517)	india (0.0334)	afghanistan (0.1023)
russia (0.0272)	iraqi (0.0566)	somali (0.0302)	indian (0.0274)	afghan (0.0760)
chechnya (0.0211)	baghdad (0.0457)	govern (0.0269)	kashmir (0.0224)	taliban (0.0676)
caucasus (0.0180)	mosul (0.0197)	mogadishu (0.0217)	pakistan (0.0180)	nato (0.0327)
kill (0.0145)	sunni (0.0151)	shabaab (0.0164)	isi (0.0147)	troop (0.0321)
chechen (0.0137)	citi (0.0143)	fight (0.0158)	mumbai (0.0121)	forc (0.0241)
oper (0.0127)	forc (0.0125)	islamist (0.0152)	attack (0.0106)	kabul (0.0197)
ingushetia (0.0104)	sourc (0.0115)	islam (0.0134)	let (0.0095)	insurg (0.0145)
north (0.0096)	aswat (0.0115)	shabab (0.0124)	arrest (0.0088)	karzai (0.0133)
moscow (0.0093)	shiit (0.0086)	town (0.0124)	lashkar (0.0087)	elect (0.0130)

Topic-based Social Network Visualization

In the following, for illustration purposes, our social network visualization and analysis will be targeted toward the "Local Conflicts" group of topics. All graphs were constructed using the Java Universal Network/Graph Framework Jung `2.0.1` and compiled with JDK `1.6.0_16`.

In Fig. 8.2, the whole corpus's social network is presented without any filtering methods. This can be interpreted as the complete network built by using the complete repliers structure, where the edge $a_{i,j}$ between user i and j is defined for every interaction between users. This

---

[5]Ansar Al-Islam (supporters or partisans of Islam) refers to a Kurdish-Sunni Islamic group, promoting a radical interpretation of Islam.

Table 8.2:   Ten most relevant words with their respective conditional probabilities for five topics associated with the "War on Terror" concept from the `Ansar1` forum

Topic 3 "War"	Topic 21 "American Soldier"	Topic 24 "Imprisonment"	Topic 25 "Obama"	Topic 31 "Military Operations"
peopl (0.0114)	islam (0.1238)	prison (0.0194)	presid (0.0146)	cia (0.0130)
war (0.0096)	emir (0.0964)	court (0.0137)	obama (0.0143)	report (0.0112)
countri (0.0092)	afghanistan (0.0811)	case (0.0104)	govern (0.0113)	million (0.0107)
american (0.0091)	provinc (0.0524)	charg (0.0099)	countri (0.0103)	work (0.0103)
world (0.0062)	destruct (0.0248)	releas (0.0092)	offici (0.0094)	compani (0.0100)
forc (0.0060)	american (0.0158)	detaine (0.0083)	secur (0.0092)	money (0.0095)
america (0.0058)	kill (0.0145)	tortur (0.0080)	militari (0.0082)	oper (0.0093)
one (0.0055)	soldier (0.0129)	alleg (0.0072)	year (0.0080)	blackwat (0.0082)
govern (0.0055)	tank (0.0104)	investig (0.0069)	minist (0.0079)	use (0.0077)
militari (0.0050)	enemi (0.0096)	guantanamo (0.0061)	state (0.0077)	state (0.0071)

Table 8.3:   Ten most relevant words with their respective conditional probabilities for five topics associated with the "Recruiting" concept from the `Ansar1` forum

Topic 0 "Religious Propaganda"	Topic 36 "Religious Conflicts"	Topic 39 "Ansar Propaganda"	Topic 40 "Family"	Topic 37 "Ideology"
islam (0.0270)	mosqu (0.0267)	ansar (0.0940)	women (0.0272)	muslim (0.0475)
movement (0.0247)	muslim (0.0254)	wmv (0.0419)	school (0.0160)	islam (0.0471)
allah (0.0217)	protest (0.0163)	info (0.0204)	children (0.0157)	law (0.0116)
youth (0.0217)	anti (0.0109)	jihad (0.0188)	year (0.0149)	rule (0.0088)
media (0.0211)	peopl (0.0096)	final (0.0184)	old (0.0149)	scholar (0.0087)
apost (0.0184)	bodi (0.0089)	ansarnet (0.0140)	woman (0.0126)	say (0.0086)
mujahideen (0.0172)	prayer (0.0085)	showthread (0.0132)	men (0.0124)	issu (0.0070)
crusad (0.0169)	ramadan (0.0077)	issu (0.0120)	man (0.0121)	group (0.0069)
god (0.0149)	christian (0.0072)	video (0.0115)	girl (0.0109)	state (0.0068)
state (0.0133)	grave (0.0069)	media (0.0086)	student (0.0103)	call (0.0065)

image represents the initial visual representation of the social network, from which it is difficult to extract useful information at first sight.

In Fig. 8.3, a filtered network is shown with only those creators whose reply posts were aligned to their respective thread topics.

Specifically for the group of topics associated with the "Local Conflicts" concept, a visualization of both the Last Reply–Oriented and Creator-Oriented Networks are presented in Fig. 8.4 and Fig. 8.5 respectively.

For the Last Reply–Oriented Network (Fig. 8.4), it is interesting to see that the interactions between the top five members as ordered by their incident replies is represented by a perfect clique

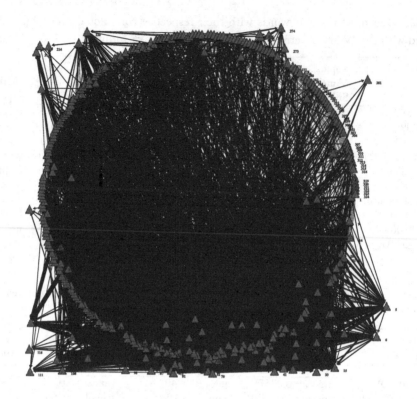

Figure 8.2: Visualization of the complete forum activity/social network representation without topic-based filtering.

structure. This could be interpreted as the following: for the "Local Conflicts" group of topics, the most replied-to members in the Ansar1 forum share common interests in this particular topic and reply to themselves in this context.

For the Creator-Oriented Network (Fig. 8.5), a clean visualization of the top eight members for the "Local Conflicts" group of topics ordered by their incident replies is presented. In this case, a first analysis leads us to see that "user-48" and "user-51" do not appear as influential thread creators, but can be considered as very important repliers in threads created by users such as "user-2," "user-3," and "user-4."

Hub-Based Social Network Analysis

As a first approach for the *topic-based community key-members extraction problem*, we propose usage of the HITS algorithm to determine those members whose posts' interactions, filtered by topics, reflect special structures and insights into whether they could be considered as key members of the topic-based social network.

Figure 8.3: Visualization of the Creator-Oriented Network for all topics during the complete period of forum activity.

The HITS algorithm [Kleinberg, 1999] has been widely used in the social network analysis community for different purposes, where the authoritative and hub scores can be interpreted as rich information on how different nodes (or community members) behave. As an example of what is the potential of the HITS algorithm for this research, in Table 8.4 the top five users are listed (with their respective hub score) for the Creator-Oriented and Last Reply–Oriented Networks for all topics. In this case, it can be seen that in neither case were the users repeated, which means that the topic-based network structure has completely different properties.

Then, in Table 8.5, the top five members are again listed but for the "Local Conflicts" topic-filtered social network. In this case, the information that this case Creator-Oriented and Last Reply-Oriented top lists highlights is that forum members are repeated, showing that this social network ("Local Conflicts") is specially built by posts of specific users (like "user-2" and "user-13").

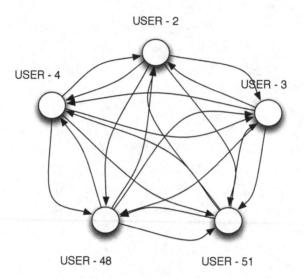

Figure 8.4: Visualization of the top five users and their interactions in the Last Reply–Oriented Network for the "Local Conflict" concept over the complete period of forum activity.

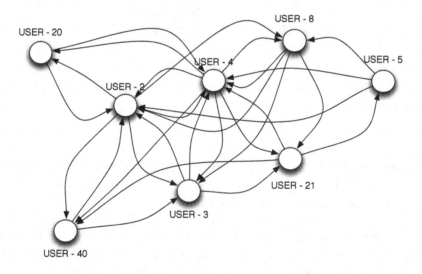

Figure 8.5: Visualization of the Creator-Oriented Network for the "Local Conflict" concept over the complete period of forum activity.

Table 8.4: Top five members ordered by HITS' hub score over the social network for the complete topics evaluation for both a Creator-Oriented Network (*type 1*) and a Last Reply–Oriented Network (*type 2*)

All topics (*type 1*)	All topics (*type 2*)
user-39 (0.07985)	user-2 (0.2666)
user-70 (0.07985)	user-4 (0.2067)
user-40 (0.07985)	user-43 (0.1965)
user-16 (0.07985)	user-13 (0.1781)
user-53 (0.07985)	user-3 (0.1753)

Table 8.5: Top five members ordered by HITS' hub score over the social network for the "Local Conflict" concept for both a Creator-Oriented Network (*type 1*) and Last Reply–Oriented Network (*type 2*)

"Local Conflicts" (*type 1*)	"Local Conflicts" (*type 2*)
user-2 (0.3757)	user-2 (0.3001)
user-13 (0.2333)	user-43 (0.2221)
user-43 (0.2125)	user-3 (0.2203)
user-25 (0.2055)	user-4 (0.2142)
user-24 (0.1937)	user-13 (0.1955)

8.5   CONCLUSION

We propose to combine traditional SNA with data mining techniques in order to produce results which enable us to measure social aspects which are not considered by applying SNA alone. In this way, we can obtain results closer to reality when performing further analysis like experts detection, sub-groups detection, centrality measures, among other measures, on any virtual community, VCoP, VCoI, or other human-based interaction systems such as social networks.

We used our approach to study a VCoI called the Dark Web, for which the Ansar1 forum was used. This forum was collected from December 2008 to January 2010, and was created by 376 members in 29,057 posts. By applying Latent Dirichlet Allocation (LDA) and using Average Shortest Distances (ASD), we obtained 47 topics using the closest 30 words for each topic. Afterward, 11 topics were discarded by inspection. Finally, we used two different topology representations for the forum: Creator-Oriented and Last Reply–Oriented Networks.

We showed in which ways SNA alone could be very poor in detecting key members who are specifically talking about certain subjects. However, when combining a topic-based text mining approach with SNA, we outperform SNA alone to discover VCoIs' key members, for which a better analysis of social networks can be performed.

In our experiments, we used 14 months of social network data which was quite dense, and performed SNA. Its visual representation is not very informative to help us identify patterns. However, using our method, we were able to focus on a specific group of topics to create a topic-based network, which provides specific information since it less dense. Then, by using HITS, key members can be extracted from both network configurations, and also the results are radically different.

The work in this chapter leads to much better results than just applying SNA on its own. However, as future work the overall evaluation of the top ranked members by the HITS algorithm can be rated by expert interviews. Also, by combining other SNA-based measures and techniques, such as betweenness centrality scores, weighted HITS, and HITS authoritative scores, the key members could be identified with better performance. Also, a simple visual inspection of the topic-based SNA can allow us to view a perfect subgroup with key members.

In this chapter, we have successfully tested our approach and we have shown how it can be applied to the Dark Web to discover potential groups of topics (e.g., potential homeland security threats), along with key members that populate these topics.

Last but not least, we can store and manage concepts/topics using SIOCM. We can also store members' own topic preferences, and we can identify those interested in the topics of other members. All this knowledge is stored in a semantic format which can later be used by other algorithms that can mesh this knowledge with other data to create an augmented dataset from the community. In this way, a knowledge discovery process can be truly reused based on a common Semantic Web data structure like SIOCM.

In the next chapter, we will describe how we can move from the meso- to micro-level, by looking more at the topics of interest to each user.

CHAPTER 9

Social Semantic Web Mining of Users

9.1 INTRODUCTION

In the past, a variety of techniques have examined the best ways to recommend content to people and provide more custom search functionality to that person based on their user account (profile) and activities on a particular website. However, a person is not solely represented by an isolated profile on a single site, but rather has a multi-faceted profile that consists of their identity and activities across a range of websites, as shown in Fig. 9.1.

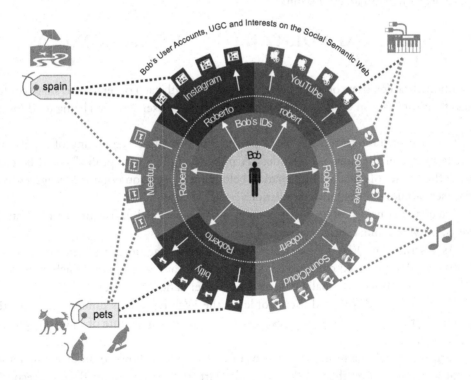

Figure 9.1: The user is more than just a single profile.

A person may use different social networks for specific purposes: staying in touch with friends and family (Facebook), for information delivery (Twitter), or business networking (LinkedIn). However, each site can use aspects of the user profile from other sites to provide better recommendations and search.

The Social Semantic Web is an enabling technology that also allows us to link a person to their various user accounts (as mentioned earlier) and to the activities carried out through those accounts.

In this chapter, based on work carried out with Orlandi and Passant [Orlandi et al., 2012], we will describe how we can use the Social Semantic Web to mine information about a particular user, and in particular, how we can create an interoperable user interest profile automatically.

We will begin by describing the process (and reasons) for creating this user interest profile, i.e., how the user profile is modeled and represented. This profile should also take into account provenance, since this is key to knowing where information was created on the Social Semantic Web.

We describe some of the heuristics that can be used for mining interests, comparing the bag-of-words approaches referred to in Chapter 2 to disambiguated entities from DBpedia. We will then show the architecture for collecting data, and creating and aggregating the multi-source user profile, along with an evaluation.

9.2   MODELING A DISTRIBUTED USER PROFILE WITH INTERESTS

As mentioned earlier, users on the Social Web interact with each other, create/share content, and express their interests on various social websites with many user accounts being used for different purposes.

On each of these systems, a user's personal information—consisting of a portion of the complete profile of the user—is recorded. A more "complete user profile" could be created by deriving the full set of personal information belonging to a person obtained by aggregating their partial user profiles distributed across various Social Web systems.

Each partial user profile might contain a user's personal and contact information, their interests, activities, and a social network of their contacts.

As described in Orlandi et al. [2012], we will look at aggregating user profiles of interests with structured and ranked collections of concepts relevant to the users. Such concepts can be used by applications for personalization and recommendation purposes.

Each part of a distributed user profile on the Web represents different *facets* of the user, and therefore their aggregation provides a more comprehensive picture of a person's profile [Abel et al., 2010a].

Aggregation of user profiles brings several advantages: it allows for information reuse across different systems; it solves the well-known "cold start" problem in personalization/recommendation systems; and it provides more complete information to each individual Social Web service.

However, the aggregation process is a non-trivial problem which involves a number of commonly occuring data integration issues: entity matching and duplicates removal, conflicts resolution, heterogeneity of models for the source data, and the consequent need for a common model for the target data.

Using standard semantic technologies to represent the data sources can help in solving these issues and it helps to provide a unified representation for the target data model. Furthermore, a complete semantic representation and management of the provenance of user data addresses some of the duplicate/conflict resolution issues, since it allows us to track the origins of the data at any point in the integration process [Hartig and Zhao, 2010].

We choose the Social Semantic Web vocabularies FOAF and SIOC, introduced in Chapter 5, for modeling a distributed user interest profile. As in Fig. 9.1, a person can be linked to multiple user accounts. The person can be modeled using the foaf:Person class, as the holder of multiple user accounts (sioc:UserAccount, a sub-class of foaf:OnlineAccount).

9.3   RELATED WORK MINING USER PROFILES ON THE SOCIAL SEMANTIC WEB

Over the past few years, the growth in the popularity of the Social Web has led to many sites collecting data on users and their behavior to provide more personalized content and recommendations. When gathering information about a user across a number of sites, a reliable user model needs to be defined. Various authors have looked at some of the resulting challenges including semantic heterogeneity in user models by Carmagnola et al. [2011] and the suitability of "strong" versus "weak" semantic technologies for user modeling tasks by Torre [2009].

Szomszor et al. [2008] have presented a method that combines user information retrieval/profiling with the Semantic Web. The authors present a method to merge a user's distributed tag clouds and build a richer profile of interests using the FOAF vocabulary, while matching concepts to Wikipedia categories. In Carmagnola [2009], the authors describe an advanced user modeling system that leverages semantic technologies. The authors use RDF for representing the user models, and reasoning capabilities are implemented on top of the user models in order to obtain an automatic mapping between heterogeneous concepts. Another extensive approach for ontology-based representation of user models was proposed by Heckmann et al. [2005]: the General User Modeling Ontology (GUMO) is a uniform interpretation of distributed user models.

For user profiling on social networks like Twitter, Tao et al. [2011] and Abel et al. [2011b] present approaches that create FOAF-based user profiles using the frequency of entities extracted from a user's tweets. The authors also show how an aggregation of a user's tags can be used to help build personal semantically enriched interest profiles. Similarly, Stan et al. [2011] describes a system for recommending people based on what they term *User Interaction Profiles*, created by extracting entities from keywords in user posts on social networks (specifically Twitter in their paper). Disambiguation and concept expansion is carried out using DBpedia and other semantic technologies.

Aroyo and Houben [2010] discuss the challenges for user modeling and personalization using adaptive Semantic Web applications, and presents a review of research in the field. It is important to note the advantages that Semantic Web–based approaches bring over some others: if reasoning is carried out on top of user data without some common semantic model, it can make interoperability with other systems more difficult to achieve.

9.4   REPRESENTING USER INTEREST PROFILES

In order to extract and generate a distributed user profile from multiple social networking websites, we do the following: *Firstly*, we extract the data from each social networking service and generate application-specific user profiles. After this, the *second* step involves representing the user models using popular ontologies, and then, *finally*, we aggregate the distributed profile.

We will now describe an RDF-based modeling solution for multi-domain user profiles with interests, and later detail how we can integrate this user data with the Web of Data and in particular DBpedia.

Semantic Web technologies and de facto standard Social Web ontologies are the main frameworks used in the Social Semantic Web for the development of interoperable services, and these standards make it easier to connect distributed user profiles. Our solution for modeling profile data is mainly based on the SIOC and FOAF vocabularies, as described in Chapter 5.

FOAF, detailed in Section 5.8.1 and one of the most popular lightweight ontologies on the Semantic Web, is used as the basis for representing a user's personal information and social relationships. As such, it can help with the integration of heterogeneous distributed user profiles. A FOAF profile consists of a FOAF `PersonalProfileDocument` that describes a `foaf:Person`: a human that has several properties describing him or her and the online accounts that he or she holds/owns on the Web.

For the purposes of our model, an important part of the user profile is how to represent the user's interests. We want to automatically retrieve a user's interests from multiple social networking sites and compute weights that can express their relevance.

In Listing 9.1 and Fig. 9.2, we display an example of an interest (a `WeightedInterest`) about the entity "*Semantic Web*" with a weight of *0.5* on a specific scale (from 0 to 1) using the *Weighted Interests Vocabulary* (WI)[1] and the *Weighting Ontology* (WO).[2]

In order to compute weights for the interests, most approaches are based on the number of occurrences of the entities, their frequency, and potentially some additional factors. These factors might depend on: whether or not the interest was implicitly mined or explicitly expressed by the user; a time-based function which computes the decay of the interests over time; the trustworthi-

---

[1]http://purl.org/ontology/wi/core (accessed September 2014), derived in part from http://rdfs.org/imo/ns by Passant, Kinsella, and Breslin.
[2]http://purl.org/ontology/wo/core (accessed September 2014).

ness of the social platform; etc. In Section 9.7, we will describe some different weighting schemes and the heuristics we used.

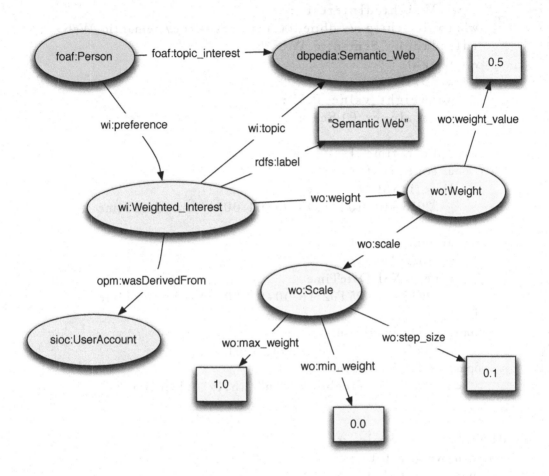

Figure 9.2: Example of our modeling solution for user interests.

Listing 9.1: An RDF/Turtle representation of an interest (Semantic Web) and its weight (0.5) extracted from two sources at two different time instants

```
@prefix ##please visit http://prefix.cc for the prefixes##

<http://example.org/fabrizio#me>
    a foaf:Person ;
    foaf:name "Fabrizio Orlandi" ;
    foaf:topic_interest
```

```
        <http://dbpedia.org/resource/Semantic_Web> ;
    wi:preference [
        a wi:WeightedInterest ;
        wi:topic <http://dbpedia.org/resource/Semantic_Web> ;
        rdfs:label "Semantic Web" ;
        wo:weight [
            a wo:Weight ;
            wo:weight_value 0.5 ;
            wo:scale ex:01Scale
            ] ;
        wi:appear_time [
            a time:Instant ;
            time:inXSDDateTime
                "2013-10-16T11:30:00+01:00"^^xsd:dateTime
        ]
        wi:appear_time [
            a time:Instant ;
            time:inXSDDateTime
                "2013-11-05T02:18:00+01:00"^^xsd:dateTime
        ]
        opm:wasDerivedFrom
            <http://twitter.com/BadmotorF> ;
        opm:wasDerivedFrom
            <http://www.facebook.com/fabriziorlandi> ;
    ] .

ex:01Scale a wo:Scale ;
    wo:min_weight 0.0 ;
    wo:max_weight 1.0 ;
    wo:step_size 0.1 .

<http://twitter.com/BadmotorF> a sioc:UserAccount .
<http://www.facebook.com/fabriziorlandi> a sioc:UserAccount .
```

We use a model based on FOAF and SIOC with the addition of other more specific vocabularies to produce a detailed description of a user's interests. Terms in the model are highly integrated; the model facilitates trivial interchange of information across heterogeneous social platforms.

In Listing 9.1, we have also used the property `appear_time` and an `Instant` class from the W3C Time Ontology[3] to describe in a generic way the particular time instant at which the interest was expressed in the user's social activity.

More detailed temporal dynamics (such as time intervals) and context (such as events and places) related to the interests can be expressed using the `InterestDynamics` class and related properties, as detailed in the WI ontology specification[4] which has more details about how to model a Social Web user's interests.

9.5   LEVERAGING THE PROVENANCE OF USER DATA

The provenance of data is quite important in the context of this interest profile as it allows data consumers to understand the origins of the interests (time- and source-wise) which are the result of an integration process.

Some data consumers might want to give more relevance to some data sources rather than others according to particular trust measures or differences in contexts and use cases. Also, it would be possible to recompute new aggregated weight values based on different weighting schemes and the original data, or to enforce privacy rules on user data based on particular preferences.

As regards provenance of the interests, as shown in Listing 9.1, we use the property `wasDerivedFrom` from the Open Provenance Model (OPM) to state that the interest originated in a specific user account on a website. This property is equivalent to the PROV `wasDerivedFrom` property. In Listing 9.1, we observe that the interest is derived from both the user's Twitter and Facebook accounts.

For a complete provenance representation of a user's interests, we can record more information about the origin of an interest. In particular, in line with the W7 model,[5] we can connect each *interest* to:

- **(Who)** The agent (a `foaf:Person`) who expressed the interest;

- **(When)** The time when it was initially expressed/created or modified (using the `wi:appear_time` property);

- **(What)** Its dereferencable description (using `foaf:topic_interest` and `wi:topic`, and pointing to a DBpedia resource);

- **(Which)** The website or user account it was extracted from (using `opm:wasDerivedFrom`);

- **(How)** The Social Web action which caused the interest to be expressed, for example, a `sioca:Action` (or alternatively `opm:Process` or `prov:Activity`) can be connected to the `WeightedInterest` through the property `sioca:creates` (you can read more about the SIOC Actions module in Section 5.8.3);

---

[3]http://www.w3.org/TR/2006/WD-owl-time-20060927/ (accessed September 2014).
[4]http://smiy.sourceforge.net/wi/spec/weightedinterests.html (accessed September 2014).
[5]http://arizona.openrepository.com/arizona/bitstream/10150/145314/1/azu_etd_11468_sip1_m.pdf.

- **(Where)** We may not have related information about a physical location in this case;

- **(Why)** We may not have precise information about the reasons behind a user having a particular interest.

9.6   INTERESTS ON THE WEB OF DATA

In order to disambiguate the interests that we extract for a user, we take the approach of linking them to an entity on the Web of Data. In this chapter, we will show how we use DBpedia (which contains semantic concepts and relationships extracted from Wikipedia) for these entities. DBpedia is a useful source of entities, containing about 1 billion RDF triples, and is one of the most important and interlinked datasets on the Web of Data [Bizer et al., 2009]. However, we could equally use any source of entities providing it had enough coverage of the domain of interests (e.g., DMOZ/Open Directory Project categories).

Representing interests using DBpedia resources has two main advantages: it integrates our user profiles with the Linked Open Data cloud, and provides a larger and "fresher" set of terms when compared to traditional taxonomies or lexical databases such as WordNet.[6] In Ponzetto and Strube [2007], the authors demonstrate the benefits of using Wikipedia (or DBpedia) for computing semantic relatedness and for named-entity representation compared to WordNet and other knowledge bases.

Later in this chapter, we will show how we can use DBpedia to not only link to its entities, but also to extract related categories for concept expansion and to analyze the structure of the categories graph in order to understand the relevance of a category for representing a user interest.

9.7   INTEREST MINING ON THE SOCIAL WEB

9.7.1   BAG-OF-WORDS VS. DISAMBIGUATED ENTITIES

Most state-of-the-art methods identifying possible concepts of interest in textual user-generated content (for building user profiles) employ *tag-based* user profiles [Abel et al., 2010b, Michlmayr et al., 2007]. Social Web textual content is analyzed and processed using traditional text-processing techniques such as stemming and stop-word removal to identify words or tags that frequently occur in the corpus. Sets of frequent words/tags ranked by their frequency are then added to the user profile. This methodology can lead to errors as it does not consider the position of the words in sentences, the language/grammar used, the context of the sentences, and other possible ambiguities.

For this reason we use *entity-based* user profiles here and compare them with tag-based ones. In order to identify entities within the text, we use specific tools that offer natural language processing capabilities and named entity extractors that will spot entities such as places, persons, organizations, etc., and provide the DBpedia resources associated with these entities. As described

---

[6]http://wordnet.princeton.edu.

later in Section 9.8, these tools perform entity disambiguation, as entities are linked to URIs on the Linked Open Data cloud, and ambiguities are resolved by analyzing the context of the sentences.

9.7.2   TIME DECAY OF INTERESTS

Interests will vary in importance or relevance for any user over time, and preferences that have been expressed by a user some time in the past will probably be less relevant than interests which have been expressed very recently. We can state that, in general, the relevance of interests for a user decays with the time. This condition has been verified in work by Ding and Li [2005], Abel et al. [2011a], and Nakatsuji and Fujiwara [2012].

As suggested by Ding and Li [2005] and most other states of the art, an exponential time decay function can be used to compute the relevance of interests over time. This type of function has been evaluated in the aforementioned work and has demonstrated its efficiency in many other profiling algorithms.

We use an exponential decay function to evaluate the relevance of each interest according to its position on the user timeline, as it has been shown to be simple and effective. The function gives higher weights for interests that have occurred recently and lower weights for older interests: we can assume that recent entities of interest extracted from the Social Web activity of a user reflect his or her current interests more than the older ones.

The exponential decay function is:

$$x(t) = x_0 e^{-t/\tau} \tag{9.1}$$

where $x(t)$ is the quantity at time t, $x_0 = x(0)$ is the initial quantity (at time $t = 0$), $\tau = 1/\lambda$ is a constant called the *mean lifetime*, and λ is a positive number called the *decay constant*. When an interest reoccurs multiple times, we use the average of the timestamps of the different reoccurring events as time t.

Applying this function to our use case, in order to compute the time decay for interests, we need to arbitrarily choose values for x_0 and τ which are constants of the function. For our purposes, we set $x_0 = 1$, the maximum possible value of the function. We also define an initial time window where the interests are not discounted by the decay function (7 days). The constant τ represents the time at which the function value is reduced to $1/e = 0.368$ times its initial value x_0 (we chose to evaluate with $\tau = 120 days$ and $\tau = 360 days$ following experiments and similar values in Abel et al. [2011a]).

The exponential decay function is directly applied to the frequency value of the interests, calculated as the ratio between the number of occurrences of the interest and the total number of occurrences of all interests. As regards the time considered for the decay function (the value of t), we compute the average time of the timestamps collected for each interest. Following the computation of the weights for all interests, all of the values are then normalized to an interval between 0 and 1.

One interesting distinction that could also be made is between long-term interests and short-term or occasional interests. Interests reoccurring many times while separated by long time periods indicate stable/long-term interests, while the opposite would happen for occasional interests. However, such a distinction is a subject for future work, and not described in this book.

9.7.3 CATEGORIES VS. RESOURCES

In Section 9.6, we mentioned that we link every entity or concept representing a user interest to the Web of Data, and in particular to DBpedia resources. DBpedia has two different types of resource: either standard resources—which correspond to entities or pages on Wikipedia—or categories—which represent groups of resources or Wikipedia articles. Therefore, we will look at two different types of user profiles that can be created: using *resource-based* and *category-based* methods.

The category-based methods extract all categories related to the DBpedia resources that have been computed with the resource-based methods. As soon as we get a DBpedia resource from the entity recognition tool, this forms part of the resource-based profile. Then, for every resource collected, we query the DBpedia SPARQL endpoint[7] to retrieve categories that are connected to these resources.

A DBpedia resource is linked to its categories through the Dublin Core[8] subject property. From each category, which is defined as a skos:Concept, it is also possible to navigate the categories graph to obtain more related categories using the skos:broader and skos:narrower relationships. This would be useful in use cases where it is necessary to broaden a user interests profile, for example, for recommendation systems.

Once all the categories have been retrieved from DBpedia, starting from the original resource-based user profile, we can create the category-based profile and assign different weights to the categories according to different weighting schemes.

The first weighting scheme propagates the weights of the resources computed with any resource-based method to the categories. Hence, the weight of each category is the sum of all weights of the interests/resources belonging to that category.

The second weighting scheme reduces the weight of the category (computed in the same way as the first weighting scheme) if the category is a "broad" or generic category (since it may not be descriptive or useful enough for inclusion in a user profile).

When analyzing the structure of categories on DBpedia, we noted that generic categories usually contain many resources or have several subcategories. We therefore implemented a solution to lower the weight of this type of category. A category discount value that is applied to the original weight of the category is computed as follows:

---

[7]http://dbpedia.org/sparql (accessed September 2014).
[8]http://dublincore.org/documents/dcmi-terms/ (accessed September 2014).

DBpedia resources	Weight
The_Clash	0.82
Alternative_rock	0.71
Semantic_Web	0.48
Social_media	0.42
Linked_Data	0.39
...	...

DBpedia categories	Weight
Buzzwords	0.48
Semantic_Web	0.87
Web_Services	0.48
World_Wide_Web	0.39
Hypermedia	0.39
...	...

Figure 9.3: Example of a possible resource-based profile (on the left) with relevance weights and a corresponding portion of a category-based profile (on the right) with recomputed weights.

$$Category discount = \frac{1}{log(|SP|)} \cdot \frac{1}{log(|SC|)} \quad (9.2)$$

where SP = *set of pages belonging to the category*, SC = *set of sub-categories.*

The number of sub-categories and pages is also retrieved using the DBpedia SPARQL endpoint. This method discounts the value of very generic categories such as *"Living People"* which are not meaningful or representative of a user's interest. At the same time, the method keeps the original weight for relevant and specific categories such as *"RDF."*

9.7.4 PROVENANCE-BASED FEATURES

As described in Chapter 5, the provenance of data can be useful in understanding the origin and the context of Social Web data. Provenance is particularly useful when evaluating (across different websites and/or datasets) the type and amount of contributions that are attributed to a particular user. For example, this would allow us to infer expertise, interests, and qualitative estimations on users' activities.

As argued in "Provenance Data in Social Media" by Barbier et al. [2013], it is important to identify provenance attributes on social media that could be vital to the task of identifying the provenance of information. Provenance attributes of a user may include name, location, gender, occupation, information content, preferences, etc. These attributes can help us to "understand" Social Web content, narrow down possibly relevant sources, and give more credibility to a piece of information. For example, Barbier et al., in Barbier and Liu [2011], show how many attributes of a user can be collected from Twitter alone.

Here, we will focus on attributes on social media identifying the potential interests of a user. We can utilize several provenance features to improve the accuracy of our interest mining. In particular, we may use several features in the extracted Social Web data for which we can evaluate the impact or influence on the accuracy of user profiles of interests.

In order to determine and score the interests of users, Orlandi [2014] studied the different types of actions that can be performed via Social Web systems. On each of these systems, personal information, consisting of a portion of the complete profile of the user, is recorded. Each partial user profile might contain the user's personal and contact information, his or her interests, activities, and social network of contacts. To create a "complete user profile," the full set of personal information belonging to that person can be gathered by aggregating their distributed partial user profiles on each Social Web system.

Orlandi [2014] looked at various Social Web activities and online user interactions (such as comments, status updates, likes, etc.) extracted from popular vocabularies on the Web (SIOC, Activity Streams, etc., as described in Sections 5.8 and 5.8.4) and mapped to popular social media sites such as Twitter, Facebook, and Wikipedia (although it can be extended to other services). Furthermore, these actions and content features were filtered to those that can be used to mine a user's interests. Provenance of these actions is recorded and the impact of different types of actions on the quality of the user profiles can be analyzed.

The resulting user profiles consist of entities and concepts potentially representing interests, activities, and contexts of the users based on their actions and content generated on the social networks. DBpedia resources can be used to represent user interests, and to determine a score that measures their prominence based on particular heuristics (see Orlandi [2014]).

Table 9.1 summarizes some of the actions and features that can be used for profiling user interests on the aforementioned three types of social media.

Table 9.1: Some actions and content features for mining user interests, which can indicate an interest "explicitly" or "implicitly"

	on Facebook	on Twitter	on Wikipedia
Implicit Interests	- comments - status updates - direct post to friends - checkins - media object actions (e.g., posting a video)	- user posts - user replies - retweets - followees' posts - favorite tweets - lists	- text edits - infobox/link edits - "Talk" page edits - article creation
Explicit Interests	- profile: education - profile: workplace - profile: interests - likes		- article creation - add to "Watchlist"

The features listed in Table 9.1 can be directly retrieved or extracted from the structured or textual content retrieved from the APIs for Facebook, MediaWiki (for Wikipedia), and Twitter. Most of these features are connected to textual information that can be analyzed using natural language processing tools in order to spot resources of interest. As shown in Table 9.1, some features *explicitly* express an interest in an entity, and others do so *implicitly*. The implicit interests carry a higher degree of uncertainty about a user's actual interest in the entity. This can be due to the way that the entities of interest are extracted (e.g., for comments and posts we need to use NLP tools), or because the social action involved does not necessarily imply an interest (e.g., Twitter lists or Wikipedia edits).

The features listed in Table 9.1 can be retrieved and semantically represented for a user's collected social data. We will later discuss how prioritizing (or giving more relevance to) certain types of actions can affect the accuracy of the profiling algorithm, since some entities extracted from those actions may lead to better or worse interests for a user profile.

Other provenance-based features that can be analyzed include: the type of social media source (whether it is a microblog or a wiki, etc.), the social media site (e.g., Facebook, Twitter, etc.), the time dimension (see Section 9.7.2 for information about how a time decay factor can be used), and whether the entities of interest are extracted implicitly or explicitly (as shown in Table 9.1).

We are now seeing some social media sites that allow users to provide information about another user's interests. For example, on LinkedIn, users can identify others as experts on particular topics (through skills or knowledge endorsements). This can be considered as explicit information about one's interests which is provided by others in a crowdsourced manner.

9.8   AN ARCHITECTURE FOR AGGREGATING USER PROFILES OF INTERESTS ON THE SOCIAL WEB

Orlandi, Breslin, and Passant [Orlandi et al., 2012] present an architecture for the automated creation and aggregation of interoperable and multi-source user profiles (Figure 9.4). A PHP-based web service asks users to log in with two of their Social Web user accounts, and returns an RDF representation of their user profile of interests. This generated interest profile is the aggregate result of their activity as analyzed from their Social Web user accounts.

From an architectural perspective, the framework is composed of three main modules (Figure 9.4):

1. Service-specific data collector;

2. Data analyzer and profile generator;

3. Profiles aggregator.

The second and third modules include the representation of the profiles of interests using the modeling solution described in Section 9.4. In the *data analyzer and profile generator* module, the semantic representation involves only one—single source—profile, while in the *profiles aggregator*, RDF is generated for the final aggregated user profile.

Service-Specific Data Collector
The first module is the one that interacts directly with the source of the profile, the social networking site. This module is responsible for the interaction with the service API, the user authentication, and the data collection from the API. In order to collect private data about users on

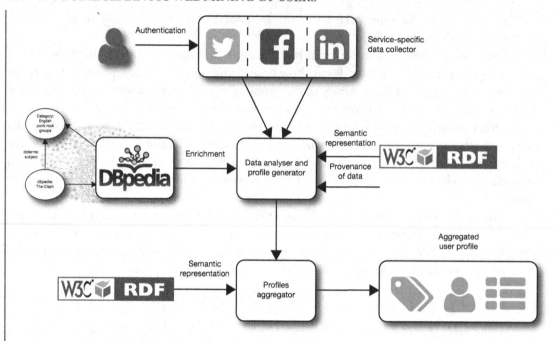

Figure 9.4: Architecture diagram from Orlandi et al. [2012].

social websites, it is necessary for the users to grant access to this data (which may be private by default).

For dealing with the chosen platforms of Facebook and Twitter, we implemented the OAuth 2.0[9] authentication system required by these platforms to access users' private data (using *Twitter-async*[10] and *Facebook PHP-SDK,*[11] respectively).

Using the Twitter API, one is able to request up to 3,200 of a user's most recent statuses, while Facebook adopts rate limits. The type of data collected from Facebook includes: status messages posted on the user's wall, the entities liked, the places checked in to, and user profile information. From Twitter, one can retrieve the status messages posted by the user on his or her timeline, and other users' messages that the user has "retweeted" or favorited.

Data Analyzer and Profile Generator
Once the user data has been collected from the different platforms, the next step is to analyze the data in order to identify entities and generate the profiles. We can use natural language processing

---

[9]http://oauth.net/2/ (accessed September 2014).
[10]https://github.com/jmathai/twitter-async (accessed September 2014).
[11]https://github.com/facebook/facebook-php-sdk (accessed September 2014).

and named entity recognition software such as AlchemyAPI,[12] AYLIEN,[13] DBpedia Spotlight,[14] Open Calais,[15] or Zemanta[16] to extract entities from the text retrieved in the previous stage.

A comparative analysis of these services is given by Rizzo and Troncy [2011] and in related research by Mendes et al. [2011]. The former says that DBpedia Spotlight and Zemanta are useful for inferring meaningful URIs, while the latter shows that Zemanta dominates in precision but has lower recall than DBpedia Spotlight.

Following the results of Rizzo and Troncy [2011] and Mendes et al. [2011], we decided to use Zemanta in combination with DBpedia Spotlight. Zemanta spots entities such as places, people, organizations, etc., and provides the related DBpedia resource for each. It performs entity disambiguation, as entities are automatically linked to URIs on the Linked Open Data cloud and ambiguities are resolved by analyzing the context of the sentences.[17] Zemanta also gives good performance when analyzing short messages such as tweets. We then use DBpedia Spotlight for when the entities detected by Zemanta have a low confidence value.

In the framework, the entity extraction algorithm is performed on every message and social activity item that has been collected during the previous stage. For each item, we then record the time that the action was performed by the user and the set of entities retrieved for that item.

A list of entities (DBpedia URIs provided by Zemanta or Spotlight) is then populated during this phase. For every entity, we record the number of occurrences and the timestamps for each of them. Therefore, it is not just the latest occurrence that is kept in memory, but also the timestamps for all previous ones (this is important for computing the weights of the interests as described in Section 9.7.2).

Finally, the set of interests generated after this second phase is represented in RDF according to the modeling solution described in Section 9.4.

Profiles Aggregator

The final phase of the framework is the aggregation of each of the user account profiles for the same person. One challenge arising when merging user profiles is the necessity to resolve shared interests reoccurring on different profiles and to recalculate a global weight for these interests. Their new aggregated weight should then be higher than the weight on a single profile, as reoccurring concepts on different social media sites indicate a strong interest. If the same interest is present on two or more profiles, it is necessary to: represent the interest only once; compute its new global weight; and update the provenance of the interest while keeping track of the sources from which the interest was derived.

---

[12]http://www.alchemyapi.com/ (accessed September 2014).
[13]http://www.aylien.com (accessed September 2014).
[14]http://dbpedia.org/spotlight (accessed September 2014).
[15]http://www.opencalais.com/ (accessed September 2014).
[16]http://www.zemanta.com/ (accessed September 2014).
[17]http://www.zemanta.com/developer/ (accessed September 2014).

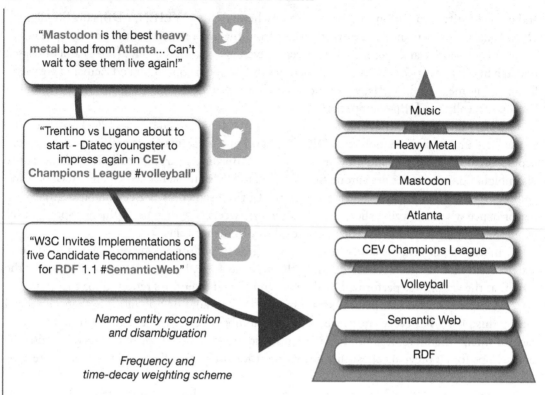

Figure 9.5: Illustrative example for interest mining from a Twitter feed of messages.

As regards the computation of the aggregated global weight, we propose a simple generic formula that can be adopted for merging the interest values from many different sources. The formula is as follows:

$$G_i = \sum_s W_s * w_{is}$$ (9.3)

where G_i = global weight for an interest i, W_s = weight associated with the source s, w_{is} = weight for the interest i due to source s.

Using this formula, it is possible to specify static weights associated with each source depending on which source we want to give more relevance to. If every social website is considered equally relevant, we multiply each of the X interest weights by a constant W_s value of $1/X$ and then sum the results. For two equal sources, the following formula summarizes the computation of a new global weight (G): $G_i = 1/2 * w_{i1} + 1/2 * w_{i2}$. It is the same as Equation 9.3 with $W_s = 1/2 \ \forall s$.

The different values associated to W_s depend on the particular source, but can also be associated to a type of source. For example, microblogging platforms (e.g., Twitter, Identica, etc.) could be associated with the same value. This fine-grained weighting strategy could be dependent on the particular applications associated with user profiles or how often they are used by the users themselves.

9.8.1   EVALUATION OF AGGREGATED USER PROFILES

An experiment was run to evaluate the accuracy of aggregated user profiles in relation to the weighting scheme and the ranking of interests. This is described in more detail in Orlandi et al. [2012], where user interest profiles were generated from users' Twitter and Facebook user accounts. For each user, the experiment was run with six different profiling algorithms: *resource*-based profiling, *category*-based profiling *first* method, and *category*-based profiling *second* method, each of them twice because two time decay parameters (360 days and 120 days) were used.

As a "*baseline*" method to compare with, a simple traditional approach of tag-based user profiling was used. It retrieves the most frequent words from a user's posts and ranks them according to their number of occurrences. Stemming is applied and stop words are also removed.

It was found that the two methods using DBpedia resources, and not the categories, performed better than the others using categories, and at the same time the results for $\tau = 360 days$ were slightly better than for $\tau = 120 days$. Therefore it seems that a longer time frame with a smoother exponential decay function better represents a user's interests. This works well where the aim of the profile is to globally represent a user's interests and contexts, but it may need to be tweaked in other applications such as news recommendation where a "fresher" and more frequently updated user profile would perform better (see for reference the work in Abel et al. [2011c]).

There are other reasons for using categories instead of resources, however. The number of categories that can be extracted for profiling a user is almost seven times larger than the number of resources. This is particularly useful in recommendation use cases, where there is a need to get as many related concepts as possible for profiling a user. According to user feedback from surveys, DBpedia resources were often revealed to be too specific and narrow and not always appropriate for representing the user's interests. On the other hand, the categories for the first method were sometimes found to be too generic (e.g., the frequently occurring "*Category:Living_People*"), and although the second category-based method is capable of removing very broad categories from the top of the interests list, it can introduce more noise. More details are available in Orlandi et al. [2012].

Orlandi [2014] further evaluated the validity of this methodology and its provenance-based features with an additional user survey. One interesting outcome of this study was in relation to the impact of the provenance of data: what are the best social features and sources of user data that one should consider for mining user interests? To answer this question, provenance information for the collected user features was gathered/stored, and the average marks given by users to the entities extracted from these features were calculated.

As one might expect, the explicitly defined interests provided better scores than implicit ones, but implicit ones are useful for extending the number and range of entities extracted. Features such as *workplace* and *education history* on a user's profile and *checkins* had high scores and were all connected to places as entities of interest. Also, it emerged that entities extracted from tweets received by *followees* were more accurate than those from the *posts* of the user themselves. The lowest accuracy was obtained from Facebook comments and posts directed at a friend's wall, most likely because of their very noisy nature.

9.9 CONCLUSION

In this chapter, we described a system for mining information about users and their interests from their content and profiles aggregated across multiple services in a Social Semantic Web. The automatic creation of interoperable and multi-domain user profiles of interests involved two essential steps: *aggregating* and *mining* user interests extracted from social data, along with related provenance information. Some advantages of entity-based user profiles of interests when compared to traditional tag-based techniques were presented, not least of which is the fact that user profiles and their entities of interest can now be connected to the Web of Data.

CHAPTER 10

Conclusions

10.1 SUMMARY

We have explored how the Semantic Web, especially Linked Data, has made it possible to link previously disconnected social datasets and services through common semantic definitions of terms (vocabularies, ontologies). Semantic entities can also be extracted from user-generated content items through web mining, NLP techniques and other named-entity recognition systems, and therefore these content items can be connected together through common semantic definitions.

In this book, we have detailed some of the current research that is being carried out to semantically represent the implicit and explicit structures in social networks, the techniques being used to elicit relevant knowledge from these structures, and we have also presented the mechanisms that can be used to intelligently mesh the semantic representations with the intelligent knowledge discovery processes.

We started by introducing the Web and a history of its origins. This was followed with a broad coverage of the main techniques that can be used to mine information from the Web and from the activities of users when surfing the Web.

We followed this up with an overview of two parallel evolutions of the Web: the Social Web, including social media sites and social networking services, and the Semantic Web, which includes the ontologies being used to add more meaning to the Web.

We then went on to describe the Social Semantic Web: an effort to bring together mutually complementary and beneficial elements of the Semantic Web and the Social Web to help each area. We can use the Semantic Web to describe people, content objects, and the connections that bind them all together so that social sites can interoperate via semantics. In the other direction, object-centered social websites can serve as rich social data sources for semantic applications.

We looked at various applications that are used to mine more information from the Social Web, at levels ranging from communities to groups to users, including community topic detection, social network analysis for groups, and user interest ranking and disambiguation.

10.2 FUTURE WORK

Social Semantic Web Mining is an exciting and promising area of current research as we have shown. In order for it to develop further, some improvements would be useful.

The first is increased dissemination and deployment of Social Semantic Web ontologies in order to increase the level and quality of social semantic data on the Web. This can include schema.org, an improved version of OGP with greater expressivity, and other continued support and evolution of RDF vocabularies like FOAF and SIOC.

The second is improved workflows and methods for transitioning social semantic data (existing and future) into knowledge, using various web and data mining techniques. We have referred to how extensions to social semantic vocabularies like SIOCM can also be useful for mining latent information from the Social Web. There are various applications for this from the micro to macro levels (emergent behavior detection, user clustering, etc.) where provenance becomes increasingly more important as we leverage data from multiple sources.

We will leave you, the reader, with some motivations for future Social Semantic Web Mining applications that we believe will not only lead to more functionality, but the concepts enabled through the resultant semantically enhanced social spaces can motivate a range of solutions with corresponding benefits, some of which are illustrated in Table 10.1.

Table 10.1: Some motivating concepts for mining the Social Semantic Web

Motivation/Obstacles	Concepts	Solutions	Benefits
Disconnected social spaces prevent knowledge transfer	Semantically integrating social spaces	Architecture of semantic participation	Completely new semantic and social search possibilities
On the Web, even structured social information often does not have clear semantics	Lightweight, standard, and interoperable descriptions of social data	Adaptive, self-organizing terminology creation	Dramatically increased availability of open, interlinked structured social information
Support for structured representation, search, authoring, is only for a small fraction of social spaces	Wiki-like authoring of structured information	Transform wikis from content bases into knowledge bases	Special interest information domains (accounting for the majority of the Web) will dramatically profit from better collaborative authoring and search
Improve semantically enhanced social spaces	Seamless integration of social spaces	Web-scale social knowledge systems	Allows sophisticated queries, search, and retrieval over several social spaces
Knowledge is isolated in social applications	Integrating other sources of knowledge	Extended social semantic mashups	Users can easily access related information in any social space

Bibliography

Abbasi, A. and Chen, H. (2005). Applying Authorship Analysis to Extremist-Group Web Forum Messages. *IEEE Intelligent Systems*, 20(5):67–75. DOI: 10.1109/MIS.2005.81. 87

Abbasi, A., Chen, H., and Salem, A. (2008). Sentiment analysis in multiple languages: Feature selection for opinion classification in Web forums. *ACM Trans. Inf. Syst.*, 26(3):1–34. DOI: 10.1145/1361684.1361685. 87

Abel, F., Gao, Q., Houben, G., and Tao, K. (2011a). Analyzing Temporal Dynamics in Twitter Profiles for Personalized Recommendations in the Social Web. In *ACM WebSci'11*, pages 1–8. DOI: 10.1145/2527031.2527040. 109

Abel, F., Gao, Q., Houben, G., and Tao, K. (2011b). Semantic Enrichment of Twitter Posts for User Profile Construction on the Social Web. In *ESWC 2011 - The Semantic Web: Research and Applications*, pages 1–15. DOI: 10.1007/978-3-642-21064-8_26. 103

Abel, F., Gao, Q., Houben, G.-J., and Tao, K. (2011c). Analyzing User Modeling on Twitter for Personalized News Recommendations. In *International Conference on User Modeling, Adaptation and Personalization (UMAP 2011)*, pages 1–12, Girona, Spain. DOI: 10.1007/978-3-642-22362-4_1. 117

Abel, F., Henze, N., Herder, E., and Krause, D. (2010a). Interweaving Public User Profiles on the Web. In *User Modeling, Adaptation, and Personalization*, pages 16–27. Springer. DOI: 10.1007/978-3-642-13470-8_4. 102

Abel, F., Henze, N., Herder, E., and Krause, D. (2010b). Linkage, aggregation, alignment and enrichment of public user profiles with Mypes. In *Proceedings of the 6th International Conference on Semantic Systems*, pages 1–8. ACM. DOI: 10.1145/1839707.1839721. 108

Alvarez, H., Ríos, S. A., Aguilera, F., Merlo, E., and Guerrero, L. (2010). Enhancing social network analysis with a concept-based text mining approach to discover key members on a virtual community of practice. In *Proceedings of the 14th International Conference on Knowledge-based and Intelligent Information and Engineering Systems: Part II*, KES'10, pages 591–600, Berlin, Heidelberg. Springer-Verlag. DOI: 10.1007/978-3-642-15390-7_61. 82, 85

Ampazis, N. and Perantonis, S. J. (2004). LSISOM - A Latent Semantic Indexing Approach to Self-Organizing Maps of Document Collections. *Neural Processing Letters*, V19(2):157–173. DOI: 10.1023/B:NEPL.0000023449.95030.8f. 27, 28, 29

Aroyo, L. and Houben, G. (2010). User modeling and adaptive Semantic Web. *Semantic Web Journal*, 1(1):105–110. DOI: 10.1007/11527886_4. 104

Artz, D. and Gil, Y. (2007). A survey of trust in computer science and the Semantic Web. *Web Semantics: Science, Services and Agents on the*, 5(2):58–71. DOI: 10.1016/j.websem.2007.03.002. 63

Baeza-Yates, R. and Ribeiro-Neto, B. (1999). *Modern Information Retrieval*. Addison-Wesley. 7, 14

Barbier, G., Feng, Z., Gundecha, P., and Liu, H. (2013). *Provenance Data in Social Media*. Morgan & Claypool. DOI: 10.2200/S00496ED1V01Y201304DMK007. 111

Barbier, G. and Liu, H. (2011). Information provenance in social media. In *SBP11 Proceedings of the 4th international conference on social computing behavioralcultural modeling and prediction*, pages 276–283. DOI: 10.1007/978-3-642-19656-0_39. 111

Berendt, B., Hotho, A., and Stumme, G. (2002). Towards Semantic Web Mining. In *ISWC '02: Proceedings of the First International Semantic Web Conference on The Semantic Web*, pages 264–278, London, UK. Springer-Verlag. DOI: 10.1007/3-540-48005-6_21. 11

Berendt, B. and Spiliopoulou, M. (2001). Analysis of navigation behavior in web sites integrating multiple information systems. *The VLDB Journal*, 9:27–75. DOI: 10.1007/s007780050083. 9, 10, 11

Berners-Lee, T., Hendler, J., and Lassila, O. (2001). The semantic web. *Scientific American*, 284(5):34–43. DOI: 10.1038/scientificamerican0501-34. 37

Berry, M. J. A. and Linoff, G. (1999). *Data Mining Techniques: For Marketing, Sales, and Customer Support*. Wiley, 1st edition edition. 10, 19, 20

Berry, M. J. A. and Linoff, G. S. (2000). *Mastering Data Mining: The Art and Science of Customer Relationship Management*. Wiley, 2nd edition edition. 10, 12, 19, 20

Bestgen, Y. (2006). Improving Text Segmentation Using Latent Semantic Analysis: A Reanalysis of Choi, Wiemer-Hastings, and Moore (2001). *Comput. Linguist.*, 32(1):5–12. DOI: 10.1162/coli.2006.32.1.5. 9, 28

Bizer, C., Auer, S., Kobilarov, G., Hellmann, S., Lehmann, J., Cyganiak, R., and Becker, C. (2009). DBpedia - A crystallization point for the Web of Data. *Web Semantics: Science, Services and Agents on the World Wide Web*, 7:154–165. DOI: 10.1016/j.websem.2009.07.002. 108

Blei, D., Ng, A., and Jordan, M. (2003). Latent dirichlet allocation. *The Journal of Machine Learning ....* 89, 90

Bourhis, A., Dubé, L., and Jacob, R. (2005). The success of virtual communities of practice: The leadership factor. *The Electronoc Journal of Knowledge Management*, 3(1):23–34. 86

Bowman, C. M., Danzig, P. B., Hardy, D. R., Manber, U., and Schwartz, M. F. (1995). The Harvest information discovery and access system. *Comput. Netw. ISDN Syst.*, 28(1-2):119–125. DOI: 10.1016/0169-7552(95)00098-5. 15

Boyd, D. and Ellison, N. B. (2008). Social Network Sites: Definition, History, and Scholarship. *Journal of Computer-Mediated Communication*, 13(1):210–230. DOI: 10.1111/j.1083-6101.2007.00393.x. 32

Bradford, R. B. (2006). Application of latent semantic indexing in generating graphs of terrorist networks. In *Proceedings of the 4th IEEE International Conference on Intelligence and Security Informatics*, ISI'06, pages 674–675, Berlin, Heidelberg. Springer-Verlag. DOI: 10.1007/11760146_84. 87

Breslin, J. G. and Decker, S. (2007). The Future of Social Networks on the Internet: The Need for Semantics. *IEEE Internet Computing*, 11(6):86–90. DOI: 10.1109/MIC.2007.138. 70

Breslin, J. G., Passant, A., and Decker, S. (2009). *The Social Semantic Web*, Springer Publishing Company, Incorporated, 2009. 45

Carmagnola, F. (2009). Handling Semantic Heterogeneity in Interoperable Distributed User Models. *Advances in Ubiquitous User Modelling*, pages 20–36. DOI: 10.1007/978-3-642-05039-8_2. 103

Carmagnola, F., Cena, F., and Gena, C. (2011). User model interoperability: a survey. *User Modeling and User-Adapted Interaction*, pages 1–47. DOI: 10.1007/s11257-011-9097-5. 103

Carroll, J., Bizer, C., Hayes, P., and Stickler, P. (2005). Named graphs, provenance and trust. In *Proceedings of the 14th international conference on World Wide Web*, pages 613–622, New York, New York, USA. ACM. DOI: 10.1145/1060745.1060835. 63

Chakrabarti, S. (2002). *Mining the Web: Discovering Knowledge from Hypertext Data*. ISBN 1-55860-754-4. Morgan Kaufmann, 1st edition edition. 2, 9, 11, 16, 20

Champin, P. and Passant, A. (2010). SIOC in Action - Representing the Dynamics of Online Communities. In *Proceedings of the 6th International Conference on Semantic Systems (I-SEMANTICS 2010)*. ACM. DOI: 10.1145/1839707.1839722. 57

Chau, R. and Yeh, C.-H. (2004). Filtering multilingual Web content using fuzzy logic and self-organizing maps. *Neural Comput. Appl.*, 13(2):140–148. DOI: 10.1007/s00521-004-0416-1. 9

Chen, L. and Chue, W. L. (2005). Using web structure and summarisation techniques for web content mining. *Inf. Process. Manage.*, 41(5):1225–1242. DOI: 10.1016/j.ipm.2005.01.005. 9

Cohen, W. W. (1995). What can we learn from the web? In *16ᵗh Int. Conf. Machine Learning (ICML99)*, pages 515–521. 19

Cooley, R., Tan, P.-N., and Srivastava, J. (1999). Discovery of Interesting Usage Patterns from Web Data. In *WEBKDD*, pages 163–182. DOI: 10.1007/978-3-642-33326-2_4. 9

Davidson, S. B. and Freire, J. (2008). Provenance and scientific workflows: challenges and opportunities. In *Proceedings of the 2008 ACM SIGMOD conference*. DOI: 10.1145/1376616.1376772. 63

Deerwester, S. C., Dumais, S. T., Landauer, T. K., Furnas, G. W., and Harshman, R. A. (1990). Indexing by Latent Semantic Analysis. *Journal of the American Society of Information Science*, 41(6):391–407. DOI: 10.1002/(SICI)1097-4571(199009)41:6%3C391::AID-ASI1%3E3.0.CO;2-9. 27, 29

Ding, Y. and Li, X. (2005). Time Weight Collaborative Filtering. In *Proceedings of the 14th ACM international conference on Information and knowledge management CIKM 05*, pages 485–492. DOI: 10.1145/1099554.1099689. 109

Duda, R. O., Hart, P. E., and Stork, D. G. (2001). *Pattern Classification*. Wiley-Interscience, second edition. 19

Eirinaki, M., Lampos, C., Paulakis, S., and Vazirgiannis, M. (2004). Web personalization integrating content semantics and navigational patterns. In *WIDM '04: Proceedings of the 6th annual ACM international workshop on Web information and data management*, pages 72–79, New York, NY, USA. ACM Press. DOI: 10.1145/1031453.1031468. 10

Eirinaki, M. and Vazirgiannis, M. (2003). Web mining for web personalization. *ACM Trans. Inter. Tech.*, 3(1):1–27. DOI: 10.1145/643477.643478. 10

Eirinaki, M., Vazirgiannis, M., and Varlamis, I. (2003). SEWeP: using site semantics and a taxonomy to enhance the Web personalization process. In *KDD '03: Proceedings of the ninth ACM SIGKDD international conference on Knowledge discovery and data mining*, pages 99–108, New York, NY, USA. ACM Press. DOI: 10.1145/956750.956765. 10

Etzioni, O. (1996). The World-Wide Web: quagmire or gold mine? *Communications of the ACM*, 39(11):65–68. DOI: 10.1145/240455.240473. 10, 11

Fernández, S., Berrueta, D., and Labra, J. E. (2007). Mailing Lists Meet The Semantic Web. In *Proceedings of the BIS 2007 Workshop on Social Aspects of the Web (SAW2007)*, volume 245 of *CEUR Workshop Proceedings*. CEUR-WS.org. 58

Franz, T. and Staab, S. (2005). SAM: Semantics Aware Instant Messaging for the Networked Semantic Desktop. In *Proceedings of the 1st Workshop on The Semantic Desktop, 4th International Semantic Web Conference*, volume 175 of *CEUR Workshop Proceedings*. CEUR-WS.org. 58

Freitag, D. (1998). Information Extraction from HTML: Application of a General Machine Learning Approach. In *AAAI/IAAI*, pages 517–523. 15

Gery, M. and Haddad, H. (2003). Evaluation of web usage mining approaches for user's next request prediction. In *WIDM '03: Proceedings of the 5th ACM international workshop on Web information and data management*, pages 74–81, New York, NY, USA. ACM Press. DOI: 10.1145/956699.956716. 9

Gil, Y., Deelman, E., Ellisman, M., Fahringer, T., Fox, G., Gannon, D., Goble, C., Livny, M., Moreau, L., and Myers, J. (2007). Examining the challenges of scientific workflows. *Computer. IEEE Computer Society*, 40(12). DOI: 10.1109/MC.2007.421. 63

Hammond, K., Burke, R., Martin, C., and Lytinen, S. (1995). FAQ finder: a case-based approach to knowledge navigation. In *CAIA '95: Proceedings of the 11th Conference on Artificial Intelligence for Applications*, page 80, Washington, DC, USA. IEEE Computer Society. DOI: 10.1109/CAIA.1995.378787. 15

Harth, A., Gassert, H., O'Murchu, I., Breslin, J. G., and Decker, S. (2005). WikiOnt: An Ontology for Describing and Exchanging Wikipedia Articles. In *Proceedings of Wikimania 2005 – The First International Wikimedia Conference*. 58

Hartig, O. and Zhao, J. (2010). Publishing and Consuming Provenance Metadata on the Web of Linked Data. In *Proceedings of 3rd Int. Provenance and Annotation Workshop*. DOI: 10.1007/978-3-642-17819-1_10. 103

Heckmann, D., Schwartz, T., Brandherm, B., Schmitz, M., and von Wilamowitz-Moellendorff, M. (2005). Gumo – the general user model ontology. In *User Modeling 2005*, Lecture Notes on Computer Science, pages 428–432. Springer Berlin / Heidelberg. DOI: 10.1007/11527886_58. 103

Himberg, J., Ahola, J., Alhoniemi, E., Vesanto, J., and Simula, O. (2001). The Self-organizing map as a tool in knowledge engineering. In Pal, N. R., editor, *Pattern Recognition in Soft Computing Paradigm*, Soft Computing. World Scientific Publishing. 19

Honkela, T., Kaski, S., Lagus, K., and Kohonen, T. (1996). Newsgroup exploration with WEB-SOM method and browsing interface. Technical report, Helsinki University of Technology, Espoo, Finland. 28

Honkela, T., Kaski, S., Lagus, K., and Kohonen, T. (1997). WEBSOM—self-organizing maps of document collections. In *Proceedings of WSOM'97, Workshop on Self-Organizing Maps, Espoo, Finland, June 4-6*, pages 310–315. Helsinki University of Technology, Neural Networks Research Centre, Espoo, Finland. DOI: 10.1016/S0925-2312(98)00039-3. 28

Jadhav, A., Purohit, H., Kapanipathi, P., Ananthram, P., Ranabahu, A., Nguyen, V., Mendes, P. N., Smith, A. G., Cooney, M., and Sheth, A. (2010). Twitris 2.0: Semantically empowered system for understanding perceptions from social data. *Semantic Web Challenge*. 35

Jin, X., Zhou, Y., and Mobasher, B. (2004). Web usage mining based on probabilistic latent semantic analysis. In *KDD '04: Proceedings of the tenth ACM SIGKDD international conference on Knowledge discovery and data mining*, pages 197–205, New York, NY, USA. ACM Press. DOI: 10.1145/1014052.1014076. 9

Jin, X., Zhou, Y., and Mobasher, B. (2005). A maximum entropy web recommendation system: combining collaborative and content features. In *KDD '05: Proceeding of the eleventh ACM SIGKDD international conference on Knowledge discovery in data mining*, pages 612–617, New York, NY, USA. ACM Press. DOI: 10.1145/1081870.1081945. 9

Johnson, W. B. and Lindenstrauss, J. (1984). Extensions of Lipshitz mapping into Hilbert space. In *Contemp. Math*, pages 189–206. DOI: 10.1090/conm/026/737400. 27, 28

Kaski, S., Honkela, T., Lagus, K., and Kohonen, T. (1996). Creating an order in digital libraries with self-organizing maps. In *Proceedings of WCNN'96, World Congress on Neural Networks, September 15-18, San Diego, California*, pages 814–817. Lawrence Erlbaum and INNS Press, Mahwah, NJ. DOI: 10.1007/978-1-4471-1599-1_119. 28

Kim, W., Jeong, O.-R., and Lee, S.-W. (2010). On social Web sites. *Information Systems*, 35(2):215–236. DOI: 10.1016/j.is.2009.08.003. 86

Kinsella, S., Murdock, V., and O'Hare, N. (2011). "i'm eating a sandwich in glasgow": Modeling locations with tweets. In *Proceedings of the 3rd International Workshop on Search and Mining User-generated Contents*, SMUC '11, pages 61–68, New York, NY, USA. ACM. DOI: 10.1145/2065023.2065039. 35

Kirk, T., Levy, A. Y., Sagiv, Y., and Srivastava, D. (1995). The Information Manifold. In Knoblock, C. and Levy, A., editors, *Information Gathering from Heterogeneous, Distributed Environments*, Stanford University, Stanford, California. 15

Kleinberg, J. M. (1999). Authoritative sources in a hyperlinked environment. *J. ACM*, 46(5):604–632. DOI: 10.1145/324133.324140. 88, 91, 97

Knorr-Cetina, K. (1997). Sociality with objects: Social relations in postsocial knowledge societies. *Theory Culture and Society*, 14(4):1–30. DOI: 10.1177/026327697014004001. 32

Kosala, R. and Blockeel, H. (2000). Web mining research: a survey. *ACM SIGKDD Explorations Newsletter*. DOI: 10.1145/360402.360406. 14

Kosonen, M. (2009). Knowledge sharing in virtual communities – a review of the empirical research. *Int. J. Web Based Communities*, 5(2):144–163. DOI: 10.1504/IJWBC.2009.023962. 85

Kushmerick, N. (1999). Gleaning the Web. *IEEE Intelligent Systems*, 14(2):20–22. DOI: 10.1109/5254.757626. 15

Kwak, H., Choi, Y., Eom, Y.-H., Jeong, H., and Moon, S. (2009). Mining communities in networks: A solution for consistency and its evaluation. In *Proceedings of the 9th ACM SIGCOMM Conference on Internet Measurement Conference*, pages 301–314, New York, NY, USA. ACM. DOI: 10.1145/1644893.1644930. 86

Kwok, C. T. and Weld, D. S. (1996). Planning to Gather Information. In *AAAI/IAAI, Vol. 1*, pages 32–39. 15

Landauer, T. K., Foltz, P. W., and Laham, D. (1998). Introduction to Latent Semantic Analysis. *Discourse Processes*, 25:259–284. DOI: 10.1080/01638539809545028. 29

L'Huillier, G., Alvarez, H., Ríos, S. A., and Aguilera, F. (2010). Topic-based social network analysis for virtual communities of interests in the dark web. *SIGKDD Explorations Newsletter*, 12(2):66–73. DOI: 10.1145/1964897.1964917. 82, 85

Liu, B. (2006). *Web Data Mining: Exploring Hyperlinks, Contents, and Usage Data (Data-Centric Systems and Applications)*. Springer-Verlag New York, Inc., Secaucus, NJ, USA. 70

Liu, B. and Chen-Chuan-Chang, K. (2004). Editorial: special issue on web content mining. *SIGKDD Explor. Newsl.*, 6(2):1–4. DOI: 10.1145/1046456.1046457. 9

Loh, S., 233, J., De Oliveira, P. M., and Gameiro, M. A. (2003). Knowledge Discovery in Texts for Constructing Decision Support Systems. *Applied Intelligence*, 18(3):357–366. DOI: 10.1023/A:1023258306854. 75

McCallum, A., Corrada-emmanuel, A., and Wang, X. (2005). Topic and role discovery in social networks. In *In IJCAI*, pages 786–791. 86

McCallum, A., Wang, X., and Corrada-Emmanuel, A. (2007). Topic and Role Discovery in Social Networks with Experiments on Enron and Academic Email. *Journal of Artificial ....* DOI: 10.1613/jair.2229. 86

Mendes, P. N., Jakob, M., García-Silva, A., and Bizer, C. (2011). DBpedia Spotlight: Shedding Light on the Web of Documents. In *7th International Conference on Semantic Systems (I-Semantics)*, pages 1–8. DOI: 10.1145/2063518.2063519. 115

Michlmayr, E., Cayzer, S., and Shabajee, P. (2007). Add-A-Tag: Learning adaptive user profiles from bookmark collections. In *1st International Conference on Weblogs and Social Media (ICWSM'2007), Boulder, Colorado (USA).* 108

Miller, G., Beckwith, R., Fellbaum, C., Gross, D., and Miller, K. (1993). Five papers on wordnet. Technical report, Princeton University. 16

Mobasher, B. (1999). A Web personalization engine based on user transaction clustering. In *In Proceedings of the 9th Workshop on Information Technologies and Systems (WITS'99),* pages 179–184. DOI: 10.1007/3-540-44463-7_15. 9, 10

Mobasher, B., Cooley, R., and Srivastava, J. (2000). Automatic personalization based on Web usage mining. *Communications of the ACM,* 43(8):142–151. DOI: 10.1145/345124.345169. 10

Mobasher, B., Dai, H., Luo, T., and Nakagawa, M. (2001). Effective personalization based on association rule discovery from web usage data. In *WIDM '01: Proceedings of the 3rd international workshop on Web information and data management,* pages 9–15, New York, NY, USA. ACM Press. DOI: 10.1145/502932.502935. 10

Mobasher, B., Jain, N., Han, E. H., and Srivastava, J. (1997). Web Mining: Patterns from WWW Transactions. Technical report, University of Minnesota. DOI: 10.1007/978-3-540-72909-9_7. 19

Moreau, L. (2010). The Foundations for Provenance on the Web. *Foundations and Trends in Web Science,* 2(2-3):99–241. DOI: 10.1561/1800000010. 62, 63

Moreau, L. and Missier, P. (2013). PROV-DM: The PROV Data Model. 62, 63

Morisseau-Leroy, N., Solomon, M. K., and Basu, J. (2001). *Oracle 8i: Programación de Componentes Java.* Mc Graw Hill. 3

Mulvenna, M. D., Anand, S. S., and Buchner, A. G. (2000). Personalization on the Net using Web mining: introduction. *Communications of the ACM,* 43(8):122–125. DOI: 10.1145/345124.345165. 10

Muslea, I., Minton, S., and Knoblock, C. (1998). Wrapper induction for semistructured, web-based information sources. In *$2^n d$ International Conference KDD and Data Mining.* 15

Muslea, I., Minton, S., and Knoblock, C. A. (2001). Hierarchical Wrapper Induction for Semistructured Information Sources. *Autonomous Agents and Multi-Agent Systems,* 4(1/2):93–114. DOI: 10.1023/A:1010022931168. 15

Nakanishi, H., Turksen, I. B., and Sugeno, M. (1993). A review and comparison of six reasoning methods. *Fuzzy Sets and Systems,* 57(3):257–294. DOI: 10.1016/0165-0114(93)90024-C. 76

Nakatsuji, M. and Fujiwara, Y. (2012). Collaborative filtering by analyzing dynamic user interests modeled by taxonomy. In *The Semantic Web – ISWC 2012*, pages 1–16. DOI: 10.1007/978-3-642-35176-1_23. 109

Ng, T. D. and Yang, C. C. (2009). A framework for harnessing public wisdom to ensure foodsafety. In *Proceedings of the 2009 IEEE International Conference on Intelligence and Security Informatics*, ISI'09, pages 185–187, Piscataway, NJ, USA. IEEE Press. DOI: 10.1109/ISI.2009.5137297. 87

Nielsen, J. (1999). User interface directions for the Web. *Communications of the ACM*. DOI: 10.1145/291469.291470. 8

Nolker, R. D. and Zhou, L. (2005). Social Computing and Weighting to Identify Member Roles in Online Communities. *Web Intelligence, IEEE / WIC / ACM International Conference on*, 0:87–93. DOI: 10.1109/WI.2005.134. 86

Norguet, J.-P., 225, E. Z., nyi, and Steinberger, R. (2006a). Improving web sites with web usage mining, web content mining, and semantic analysis. *In Proc. SOFSEM 2006: Theory and practice of Computer Science. Lecture Notes in Computer Science*, 3831:430–439. DOI: 10.1007/11611257_41. 10

Norguet, J.-P., 225, E. Z., nyi, and Steinberger, R. (2006b). Semantic analysis of web site audience. In *SAC '06: Proceedings of the 2006 ACM symposium on Applied computing*, pages 525–529, New York, NY, USA. ACM Press. DOI: 10.1145/1141277.1141401. 9

O'Reilly, T. (2005). What Is Web 2.0: Design Patterns and Business Models for the Next Generation of Software. 6, 31

Orlandi, F. (2014). *Profiling user interests on the social semantic web*. PhD thesis, National University of Ireland Galway. 111, 112, 117

Orlandi, F., Breslin, J. G., and Passant, A. (2012). Aggregated, interoperable and multi-domain user profiles for the social web. In *I-SEMANTICS*. DOI: 10.1145/2362499.2362506. 102, 113, 114, 117

Orlandi, F. and Passant, A. (2009). Enabling cross-wikis integration by extending the SIOC ontology. In *Proceedings of the Fourth Workshop on Semantic Wikis (SemWiki2009)*. 58

Orlandi, F. and Passant, A. (2011). Modelling provenance of DBpedia resources using Wikipedia contributions. *Web Semantics: Science, Services and Agents on the World Wide Web*, 9(2):149–164. DOI: 10.1016/j.websem.2011.03.002. 63

Osterfeld, F., Kiesel, M., and Schwarz, S. (2005). Nabu – a semantic archive for xmpp instant messaging. In *Proceedings of the 1st Workshop on The Semantic Desktop, 4th International Semantic Web Conference*, volume 175 of *CEUR Workshop Proceedings*. CEUR-WS.org. 58

Pal, S. K., Talwar, V., and Mitra, P. (2002). Web mining in soft computing framework: relevance, state of the art and future directions. *Neural Networks*, 13(5):1163–1177. DOI: 10.1109/TNN.2002.1031947. 7, 8, 11, 14, 15, 19

Pathak, A. B. N. and Erickson, K. (2008). Social topic models for community extraction. In *The 2nd SNA-KDD Workshop '08 (SNA-KDD'08)*. 86

Perkowitz, M. (2001). *Adaptative Web Sites: Cluster Mining and Conceptual Clustering for Index Page Synthesis*. PhD thesis, Univerity of Washingtong, Engineering Library Building Campus Map Box 352170 University of Washington Seattle, WA 98195-2170. 10

Pfeil, U. and Zaphiris, P. (2009). Investigating social network patterns within an empathic online community for older people. *Computers in Human Behavior*, 25(5):1139–1155. DOI: 10.1016/j.chb.2009.05.001. 86

Phang, X. H. and Nguyen, C. T. (2008). GibbsLDA++. 90

Pitkow, J. (1997). In search of reliable usage data on the WWW. In *Selected papers from the sixth international conference on World Wide Web*, pages 1343–1355, Essex, UK. Elsevier Science Publishers Ltd. DOI: 10.1016/S0169-7552(97)00021-4. 10, 19

Ponzetto, S. P. and Strube, M. (2007). Knowledge derived from Wikipedia for computing semantic relatedness. *Journal of Artificial Intelligence Research*, 30:181–212. DOI: 10.1613/jair.2308. 108

Porter, C. E. (2004). A Typology of Virtual Communities: A Multi-Disciplinary Foundation for Future Research. *Journal of Computer Mediated Communication*, 10(1):00. DOI: 10.1111/j.1083-6101.2004.tb00228.x. 85

Porter, M. F. (1980). An algorithm for suffix stripping. *Program; automated library and information systems*, 14(3):130–137. DOI: 10.1108/eb046814. 16

Preece, J. (2004). Etiquette, Empathy and Trust in Communities of Practice: Stepping-Stones to Social Capital. *Journal of Universal Computer Science*. 74

Probst, G. and Borzillo, S. (2008). Why communities of practice succeed and why they fail. *European Management Journal*, 26(5):335–347. DOI: 10.1016/j.emj.2008.05.003. 86

Rehatschek, H. and Hausenblas, M. (2007). Enhancing the Exploration of Mailing List Archives Through Making Semantics Explicit. In *Semantic Web Challenge 2007, collocated with the 6th International Semantic Web Conference (ISWC)*. 58

Reid, E., Qin, J., Zhou, Y., Lai, G., Sageman, M., Weimann, G., and Chen, H. (2005). Collecting and analyzing the presence of terrorists on the web: A case study of jihad websites. In Kantor, P., Muresan, G., Roberts, F., Zeng, D., Wang, F.-Y., Chen, H., and Merkle, R., editors, *Intelligence and Security Informatics*, volume 3495 of *Lecture Notes in Computer Science*, pages 402–411. Springer Berlin Heidelberg. 87

Richardson, M., Prakash, A., and Brill, E. (2006). Beyond PageRank: machine learning for static ranking. In *WWW '06: Proceedings of the 15th international conference on World Wide Web*, pages 707–715, New York, NY, USA. ACM Press. DOI: 10.1145/1135777.1135881. 9

Ríos, S. A. (2007). A Study on Web Mining Techniques for Off-Line Enhancements of Web Sites. *Ph.D Thesis*, page 231. 75

Ríos, S. A. (2008). Semantic Web Usage Mining by a Concept-Based Approach for Off-line Web Site Enhancements. *Web Intelligence and Intelligent Agent Technology, 2008 IEEE/WIC/ACM International Conference on*, 1:234–241. DOI: 10.1109/WIIAT.2008.406. 74

Ríos, S. A., Aguilera, F., and Guerrero, L. (2009). Virtual communities of practice's purpose evolution analysis using a concept-based mining approach. *Knowledge-Based Intelligent Information and Engineering Systems - Part II; Lecture Notes in Computer Science*, 5712:480–489. DOI: 10.1007/978-3-642-04592-9_60. 73, 86, 87

Ríos, S. A. and Muñoz, R. (2012). Dark web portal overlapping community detection based on topic models. In *Proceedings of the ACM SIGKDD Workshop on Intelligence and Security Informatics*, ISI-KDD '12, pages 2:1–2:7, New York, NY, USA. ACM. DOI: 10.1145/2331791.2331793. 85

Ríos, S. A. and Muñoz, R. (2014). Content patterns in topic-based overlapping communities. *The Scientific World Journal*. DOI: 10.1155/2014/105428. 85

Ríos, S. A., Velásquez, J. D., Vera, E. S., Yasuda, H., and Aoki, T. (2005a). Using SOFM to Improve Web Site Text Content. In *Advances in Natural Computation: 1ˢt Int.l Conf., ICNC 2005*, pages 622–626, Changsha, China. Springer-Verlag GmbH. DOI: 10.1007/11539117_88. 9, 11

Ríos, S. A., Velásquez, J. D., Vera, E. S., Yasuda, H., and Aoki, T. (2005b). Web Site Improvements Based on Representative Pages Identification. In *AI 2005: Advances in Artificial Intelligence: 18th Australian Joint Conference on Artificial Intelligence*, pages 1162–1166, Sydney, Australia. DOI: 10.1007/11589990_162. 10

Ríos, S. A., Velásquez, J. D., Yasuda, H., and Aoki, T. (2006). Web Site Off-Line Structure Reconfiguration: A Web User Browsing Analysis. *Lecture Notes in Computer Science: Knowledge-Based Intelligent Information and Engineering Systems*, pages 371–378. DOI: 10.1007/11893004_48. 10

Ríos, S. A., Velásquez, J. D., Yasuda, H., and Aoki, T. (2006a). Conceptual Classification to Improve a Web Site Content. *Lecture Notes In Computer Science, Springer-Verlag*, 4224:869–877. DOI: 10.1007/11875581_104. 27

Ríos, S. A., Velásquez, J. D., Yasuda, H., and Aoki, T. (2006b). Improving Web Site Content Using a Concept-based Knowledge Discovery Process. In *IEEE/WIC/ACM Int. Conf. on Web Intelligence and Intelligent Agent Technology*, pages 361–365, Hong Kong. IEEE Computer Scociety. DOI: 10.1109/WI.2006.98. 27

Ríos, S. A., Velásquez, J. D., Yasuda, H., and Aoki, T. (2006c). Using a Self Organizing Feature Map for Extracting Representative Web Pages from a Web Site. *International Journal of Computational Intelligence Research (IJCIR)*, 2(2):159–167. 74

Rizzo, G. and Troncy, R. (2011). NERD: Evaluating Named Entity Recognition Tools in the Web of Data. In *ISWC'11 - Workshop on Web Scale Knowledge Extraction (WEKEX'11)*, Bonn, Germany. 115

Saltón, G., Wong, A., and Yang, C. S. (1975). A vector space model for automatic indexing. *Communications of the ACM archive*, Vol. 18(11):613–620. DOI: 10.1145/361219.361220. 7, 16, 88

Schocken, S. and Hummel, R. A. (1993). On the use of the Dempster Shafer model in information indexing and retrieval applications. *International Journal of Man–Machine Studies*, 39(5):843–879. DOI: 10.1006/imms.1993.1086. 7

Shummer, T. (2004). Patterns for building communities in collaborative systems. *Proceedings of the 9th European Conference on Pattern Languages and Programs*. 73

Soderland, S. (1999). Learning Information Extraction Rules for Semi-Structured and Free Text. *Machine Learning*, 34(1-3):233–272. DOI: 10.1023/A:1007562322031. 9

Spertus, E. (1997). ParaSite: mining structural information on the Web. In *Selected papers from the sixth international conference on World Wide Web*, pages 1205–1215, Essex, UK. Elsevier Science Publishers Ltd. DOI: 10.1016/S0169-7552(97)00033-0. 15

Spiliopoulo, M. (1999). Managing Interesting Rules in Sequence Mining. In *PKDD '99: Proceedings of the Third European Conference on Principles of Data Mining and Knowledge Discovery*, pages 554–560, London, UK. Springer-Verlag. DOI: 10.1007/978-3-540-48247-5_73. 9, 10

Spiliopoulou, M. (2000). Web usage mining for Web site evaluation. *Communications of the ACM*, 43(8):127–134. DOI: 10.1145/345124.345167. 9

Spiliopoulou, M., Mobasher, B., Berendt, B., and Nakagawa, M. (2003). A Framework for the Evaluation of Session Reconstruction Heuristics in Web-Usage Analysis. *INFORMS J. on Computing*, 15(2):171–190. DOI: 10.1287/ijoc.15.2.171.14445. 9

Spiliopoulou, M., Pohle, C., and Faulstich, L. (1999). Improving the Effectiveness of a Web Site with Web Usage Mining. In *WEBKDD*, pages 142–162. DOI: 10.1007/3-540-44934-5_9. 9

Srivastava, J., Cooley, R., Deshpande, M., and Tan, P.-N. (2000). Web usage mining: discovery and applications of usage patterns from Web data. *SIGKDD Explor. Newsl.*, 1(2):12–23. DOI: 10.1145/846183.846188. 9

Stan, J., Maret, P., and Do, V. (2011). Semantic User Interaction Profiles for Better People Recommendation. *International Conference on Advances in Social Networks Analysis and Mining*. DOI: 10.1109/ASONAM.2011.21. 103

Szomszor, M., Alani, H., Cantador, I., O'Hara, K., and Shadbolt, N. (2008). Semantic modelling of user interests based on cross-folksonomy analysis. *The Semantic Web-ISWC 2008*, pages 632–648. DOI: 10.1007/978-3-540-88564-1_40. 103

Tao, K., Abel, F., Gao, Q., and Houben, G. (2011). TUMS: Twitter-based User Modeling Service. In *International Workshop on User Profile Data on the Social Semantic Web (UWeb), co-located with Extended Semantic Web Conference (ESWC), Heraklion, Greece*, pages 1–15. DOI: 10.1007/978-3-642-25953-1_22. 103

Tapscott, D. and Williams, A. D. (2006). *Wikinomics: How Mass Collaboration Changes Everything*, volume 58. Portfolio. 35

Theodoris, S. and Koutroumbas, K. (2003). *Pattern Recognition*. Academic Press, second edition. 11

Torre, I. (2009). Adaptive systems in the era of the semantic and social web, a survey. *User Modeling and User-Adapted Interaction*, 19(5):433–486. DOI: 10.1007/s11257-009-9067-3. 103

Turney, P. D. (1993a). Exploiting Context When Learning to Classify. In *ECML '93: Proceedings of the European Conference on Machine Learning*, pages 402–407, London, UK. Springer-Verlag. DOI: 10.1007/3-540-56602-3_158. 9

Turney, P. D. (1993b). Robust classification with context-sensitive features. In *IEA/AIE'93: Proceedings of the 6th international conference on Industrial and engineering applications of artificial intelligence and expert systems*, pages 268–276. Gordon and Breach Science Publishers. 7

Turney, P. D. (2000). Learning Algorithms for Keyphrase Extraction. *Inf. Retr.*, 2(4):303–336. DOI: 10.1023/A:1009976227802. 7

Turney, P. D. (2001). Thumbs up or thumbs down?: semantic orientation applied to unsupervised classification of reviews. In *ACL '02: Proceedings of the 40th Annual Meeting on Association for Computational Linguistics*, pages 417–424, Morristown, NJ, USA. Association for Computational Linguistics. DOI: 10.3115/1073083.1073153. 9

Turney, P. D. (2002). Mining the Web for Lexical Knowledge to Improve Keyphrase Extraction: Learning from Labeled and Unlabeled Data. Technical report, National Research Council of Canada. 9

Turney, P. D. (2003). Coherent keyphrase extraction via web mining. *In Procs. of the 18^{t}h Int. Conference on Artificial Intelligence (IJCAI-03)*, pages 434–439. 7

Turney, P. D. and Littman, M. L. (2005). Corpus-based Learning of Analogies and Semantic Relations. *Mach. Learn.*, 60(1-3):251–278. DOI: 10.1007/s10994-005-0913-1. 9

Völkel, M. and Oren, E. (2006). Towards a Wiki Interchange Format (WIF) - Opening Semantic Wiki Content and Metadata. In *Proceedings of the First Workshop on Semantic Wikis - From Wiki to Semantics (SemWiki-2006)*, volume 206 of *CEUR Workshop Proceedings*. CEUR-WS.org. 58

Wasserman, S. and Faust, K. (1994). Social network analysis: Methods and applications. *books.google.com*. 86

Wenger, E., McDermott, R., and Snyder, W. (2002). Cultivating Communities of Practice: A Guide to Managing Knowledge. *Harvard Business School Press*. 73

Widom, J. (2005). Trio: A system for integrated management of data, accuracy, and lineage. In *CIDR*. 65

Xing, D. and Girolami, M. (2007). Employing Latent Dirichlet Allocation for fraud detection in telecommunications. *Pattern Recognition Letters*, Vol. 28(13):1727–1734. DOI: 10.1016/j.patrec.2007.04.015. 90

Xu, J. and Chen, H. (2005). CrimeNet explorer: a framework for criminal network knowledge discovery. *ACM Transactions on Information Systems (TOIS)*. DOI: 10.1145/1059981.1059984. 87

Xu, J. and Chen, H. (2008). The topology of dark networks. *Commun. ACM*, 51(10):58–65. DOI: 10.1145/1400181.1400198. 87

Xu, J., Chen, H., Zhou, Y., and Qin, J. (2006). On the topology of the dark web of terrorist groups. In Mehrotra, S., Zeng, D., Chen, H., Thuraisingham, B., and Wang, F.-Y., editors, *Intelligence and Security Informatics*, volume 3975 of *Lecture Notes in Computer Science*, pages 367–376. Springer Berlin Heidelberg. 87

Yang, L., Liu, F., Kizza, J., and Ege, R. (2009). Discovering topics from dark websites. *Computational Intelligence in Cyber Security, 2009. CICS '09. IEEE Symposium on*, pages 175–179. DOI: 10.1109/CICYBS.2009.4925106. 92, 93

Yelupula, K. and Ramaswamy, S. (2008). Social network analysis for email classification. In *Proceedings of the 46th Annual Southeast Regional Conference on XX*, ACM-SE 46, pages 469–474, New York, NY, USA. ACM. DOI: 10.1145/1593105.1593229. 86

Zhang, Y., Zeng, S., Fan, L., Dang, Y., Larson, C. A., and Chen, H. (2009). Dark web forums portal: Searching and analyzing jihadist forums. In *Proceedings of the 2009 IEEE International Conference on Intelligence and Security Informatics*, ISI'09, pages 71–76, Piscataway, NJ, USA. IEEE Press. DOI: 10.1109/ISI.2009.5137274. 87

Zhou, H., Zeng, D., and Zhang, C. (2009). Finding leaders from opinion networks. In *Intelligence and Security Informatics, 2009. ISI '09. IEEE International Conference on*, pages 266–268. DOI: 10.1109/ISI.2009.5137323. 86

Zhou, Y., Reid, E., Qin, J., Chen, H., and Lai, G. (2005). US Domestic Extremist Groups on the Web: Link and Content Analysis. *IEEE Intelligent Systems*, 20(5):44–51. DOI: 10.1109/MIS.2005.96. 87

Authors' Biographies

TOPE OMITOLA

Tope Omitola is a Senior Research Fellow with the Web and Internet Science Group at the School of Electronics and Computer Science in the University of Southampton. His research interests include provenance for Linked Data, semantic social data mining and text extraction, semantic search and information discovery, and semantic dataset discovery and ranking. He was educated at King's College London and at Jesus College Cambridge. He has worked in a variety of programming and software engineering positions with companies such as ARM and BT Labs. He co-authored the paper "Put In Your Postcode, Out Comes the Data: A Case Study" which won the best paper award at ESWC 2010, and has published at various high-profile conferences. Dr. Omitola is a member of AAAS, ACM and IEEE.

SEBASTIÁN A. RÍOS

Sebastián A. Ríos is an Assistant Professor in the Industrial Engineering Department of the University of Chile, where he is also Director of the Masters in Business Engineering. He is the Director of the CEINE Business Intelligence (BI) Research Center at the University of Chile, a collaborative applied research effort with Telefonica Chile. His interests include BI, the Semantic Web, social networks, latent semantics, process mining and business process redesign, and he has received funding from CONICYT and CORFO. He completed his Ph.D. in Knowledge Engineering at the University of Tokyo in 2007 after receiving the Mobukagakusho Scholarship from the Japanese Government, and has a degree in industrial engineering and IT/-data mining from the University of Chile. Dr. Ríos was an overall winner in the 1st International Competition on Plagiarism Detection at PAN 2011, and won the best invited session award at KES 2009.

JOHN G. BRESLIN

John G. Breslin is a Senior Lecturer in Electronic Engineering at NUI Galway. He is also a Research Leader at the Insight Centre for Data Analytics at NUI Galway (formerly DERI). He has been PI/budget holder for funding totaling €1.75M, was co-PI on the €15M Líon 2 DERI CSET, and led the Eurapp study of the EU app economy for the EC. He created the SIOC framework, implemented in hundreds of applications (by Yahoo, Boeing, Vodafone, etc.) on over 25,000 websites. He has written 150 peer-reviewed publications and co-authored the book *The Social Semantic Web*. He is co-founder of boards.ie, adverts.ie, and StreamGlider, and is an advisor to various tech startups. Dr. Breslin has won various best paper awards (ICE-GOV, ESWC, PELS) and two IIA Net Visionary Awards. He is Vice Chair of IFIP Working Group 12.7 on Social Networking Semantics and Collective Intelligence.

Printed in the United States
by Baker & Taylor Publisher Services

Printed in the United States
by Baker & Taylor Publisher Services